THE AMATEUR RADIO VERTICAL ANTENNA HANDBOOK

CAPT. PAUL H. LEE, USN(RET), N6PL

CQ Communications, Inc.

Second Edition 1984
Second Printing 1996

Library of Congress Catalog Card Number 74-83411
ISBN 0-943016-14-2

Published by CQ Communications, Inc.
76 North Broadway
Hickville, New York 11801 USA

Printed in the United States of America.

INTRODUCTION

When the publisher of CQ asked me to go back through the past twenty-two years and re-edit all the material I have written on vertical antennas for the publication of this handbook, I was very pleased. I have long felt that the subject of vertical antennas is one that has been given too little attention in the various antenna handbooks written for radio amateurs. Yagi arrays and cubical quads have been covered in considerable detail in specialized handbooks by other authors, but verticals have been given only small coverage in the general antenna handbooks, and many radio amateurs feel that verticals are not very useful at amateur frequencies.

Every type of antenna has its rightful place in communications system design, whether the system be commercial, military or amateur. An antenna design which is not optimum for one purpose may be just right for another use, under a different set of conditions. Therefore it behooves us all to acquire as much knowledge of the subject as possible, and to apply that knowledge to the solution of our communications problems.

This handbook is my attempt to spread some knowledge around, primarily among radio amateurs. The subject of antennas is one which has been near and dear to me ever since I began in amateur radio in 1931, and this interest has grown throughout my amateur work, my naval communications engineering, and my commercial radio engineering endeavors. It has been my good fortune to do considerable theoretical design, as well as to see the theory substantiated in actual practice by myself and by others. There is a wealth of material available in textbooks, research reports, technical magazines and journals, and engineering papers on the subject of antennas. In writing my articles and in the preparation of this book, I use such references liberally, giving credit to others where credit is due, and thus providing the reader with additional sources of information which he can use if he has the interest and initiative to do so.

Capt. Paul H. Lee, USN(Ret), N6PL

DEDICATION

I should like to dedicate this book to my family whose patience and forebearance with my interests in amateur radio has been great and to my friends and associates throughout my years in amateur, naval and commercial communications who have helped me acquire the knowledge and experience on which this work is based. They are too numerous to mention by name, but they will read this and understand my grateful feelings.

Captain Paul H. Lee, USN(Ret), N6PL

Edited by Alan M. Dorhoffer, K2EEK

Artwork by K&S Graphics

TABLE OF CONTENTS

OPTIMUM ANTENNA DESIGN FOR DX

In order to work DX effectively the vertical angle of radiation must be as low as possible. This chapter discusses the effects of antenna types and height upon this angle of radiation.

ONE of the most important factors in the planning of a high frequency communications system is antenna design. For a specific point-to-point circuit, both receiving and transmitting antennas are optimized as to forward gain, front-to-back ratio, vertical angle of radiation, vertical pattern shape, and available space. This process usually results in a large antenna which points in a fixed direction and which has a large number of elements or large aperture area.

For the average amateur, the process of antenna design for the higher frequencies, if it can be so called, usually consists of buying a prefabricated Yagi array, mounting it on a short tower or on his house top, and going on the air. Those who prefer the lower frequency bands, on the other hand, may end up with a random length of wire, loaded up by means of a matching network and at some height above ground which is usually far from optimum. When one considers the lack of thought which is given to such matters, or the surroundings which may consist of trees, other houses, or power lines, it is sometimes amazing that signals get out of the backyard at all.

The field of propagation and antenna design is a most fascinating subject, and one which has so many ramifications that it would be impossible to cover all of them, even in a book many times this size. This chapter is directed to those who are interested in optimizing an antenna design for amateur DX work. It is here that proper antenna design can really pay off.

Vertical Radiation Angle

Take a long look at fig. 1. This graph shows the relation of vertical angle of radiation to the distance to the first reflection point via F-Layer ionospheric reflection. The height of the F-layer varies between approximate limits of 200 and 350 kilometers, under day-night conditions. Pick out the vertical angle of 20 degrees above the horizon, for example, and you will note that the first reflection zone will be somewhere in the neighborhood of 600 to 900 miles. A vertical angle of 50 degrees will give a first reflection zone of 200 to 300 miles. On the other hand, a vertical angle of 5 degrees will give a first reflection zone between 1400 and 1900 miles. And an even lower angle of 3 degrees, if obtainable, will move the first reflection zone out to 1600 to 2100 miles! Do you

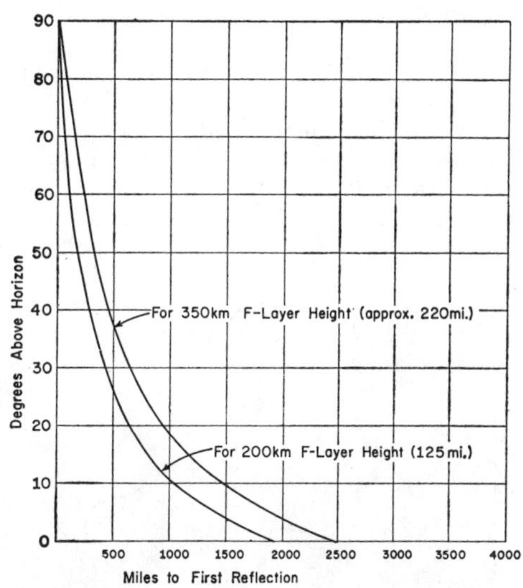

Fig. 1 — Vertical radiation angle versus distance to the first reflection.

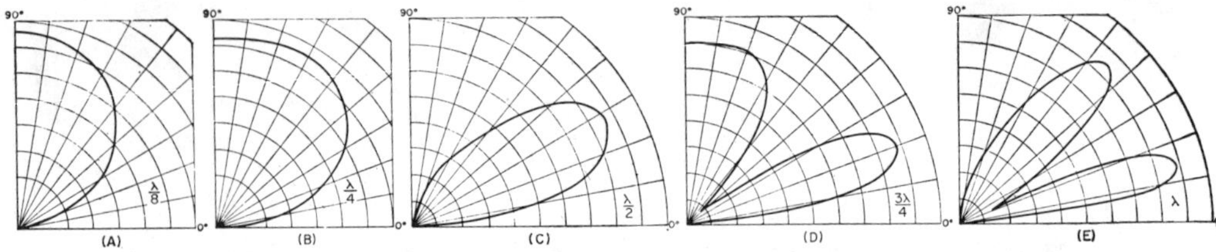

Fig. 2 — Vertical radiation patterns for half wavelength dipoles various heights above ground as indicated in each graph.

want to work DX? Then get that vertical angle down!

The next thought that comes to mind is that for long distances there will be more than one reflection. This is certainly true, for to get from Washington, D.C., to Australia, for example, will require several reflections. Due to the simple geometry of the situation, we are limited to 2500 miles per hop. However, reflection losses run in the neighborhood of 3 to 6 db per reflection, depending on the absorptive and scattering nature of the reflecting surface. When one also adds in the ionospheric absorption which occurs for each ionospheric reflection (and this absorption can be quite variable with sunspot conditions), one can easily arrive at the conclusion that for a good strong signal at an overseas receiver, the fewer the reflections the better. In other words, get that vertical angle down! Vertical angles of 3 to 10 degrees are most desirable. Curtain antenna arrays using 4 × 4 or 4 × 6 arrangements of dipole elements, such as are used by the Voice of America or the British Broadcasting Corporation, are optimized for low angles of this kind, in order to place the first reflection point as far out as possible.

How does the antenna used by Mr. Average Amateur compare with these optimum requirements? To answer this question, I have shown in fig. 2, A through E, the vertical radiation patterns of a ½ wave horizontal dipole for several heights above ground of average conductivity. The patterns are in a plane at right angles to the dipole. Let's look at some common backyard cases. Consider the case of a ½ wave dipole ½ wave high, shown in fig. 2C. This corresponds to a 33 foot dipole about 35 feet above ground at 14 mc. Note that the useful lobe is between approximately 15 and 45 degrees. Reference to fig. 1 will show that this lobe will give a first reflection zone of 200 to 1100 miles. Transmitted power is thus being scattered over quite a wide zone. Let's now think of the fellow who has a 3.9 mc dipole, hung between a couple of trees, perhaps, at a height of 30 to 50 feet. Note from figs. 2A and B that dipole heights of ⅛ to ¼ wave give a useful vertical angle of about 30 up to 90 degrees. Did I say "useful?" This type of vertical pattern is useful only for distances out to about 500 to 600 miles, and even so, a lot of power is being wasted at very high angles which penetrate the ionosphere and are not reflected at all! It looks as though we should move our dipole higher, doesn't it? Well, let's take the case of a dipole ¾ wave high. This corresponds to a height of 180 feet at 3.9 mHz, or 50 feet at 14 mHz. Figure 2D shows that we now have a lobe of radiation from 10 to 30 degrees, which is an improvement, but a high angle lobe of 55 to 90 degrees has appeared, which wastes considerable power at angles which are not useful and which probably penetrate the ionosphere and are lost. If we move the 14 mHz dipole up to a wave length height, 66 feet on 14 mHz, our low angle lobe sharpens, from about 10 to 20 degrees now, but we still have a large lobe at 38 to 60 degrees which is a waste of power. If we move our 14 mHz dipole even higher, the low angle lobe will shrink in size, several high angle lobes will appear, and our vertical pattern be-

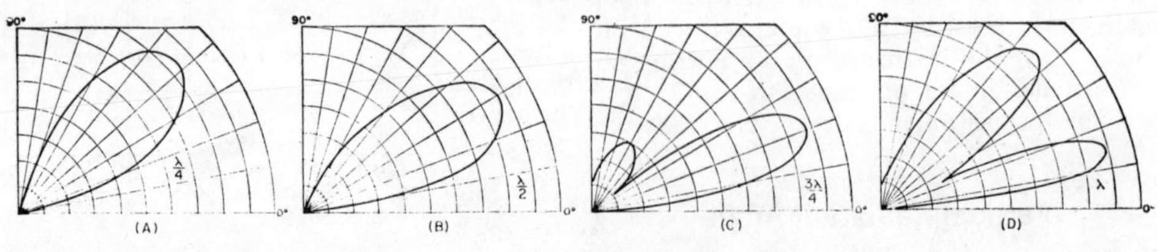

Fig. 3 — Vertical radiation patterns for half wavelength yagi antennas various heights above ground as indicated in each graph.

comes quite useless for any DX work.

Yagi Antennas

The next thought that occurs is, "Can't we improve this by using a Yagi array?" The answer to that is that a horizontal Yagi will cut down the size of the undesired high angles lobes somewhat, but that it will do absolutely nothing at all to lower the angle of the desired main lobe. Figures 3A through D show the vertical patterns of a Yagi array at various heights above ground of average conductivity. It will be noted that these figures correspond, case for case, with those pertaining to the single horizontal dipole. With a 14 mHz Yagi which is 66 feet high, we are still wasting considerable power at a useless angle of 38 to 60 degrees, and the vertical angle of the main lobe is still from 10 to 20 degrees.

How can we get lower angles than this? There is only one way to do it with horizontal dipoles, and that is by stacking them vertically. One horizontal dipole ½ wave high, with another ½ wave above it and in phase with it, will give a useful lobe from 8 to 28 degrees. Stacking a third in-phase dipole at an additional height of ½ wave, will bring the lobe down to 5 to 15 degrees. Four of them stacked will bring the lobe down to 3 to 12 degrees. Perhaps these statements will inspire some rugged individual to stack two or three 3-element full-size Yagis on 14 mHz! This is something which is done quite often by the v.h.f.-u.h.f. fraternity at frequencies where size is reasonable, but personally I don't care to do it at 14 mHz!

Verticals

There is another way of getting low angles, and this is by using a vertical antenna. Let's look at figs. 4A through D for awhile. Here are plotted the vertical patterns of vertical antennas of several heights when worked against ground of average conductivity. The ¼ wave vertical gives a useful vertical lobe from about 10 to 55 degrees. A ⅜ wave vertical will drop the lobe somewhat, from 8 to 40 degrees. A ½ wave vertical is quite an improvement, giving a useful lobe from about 5 to 35 degrees. The ⅝ wave vertical, which is the optimum height used by many broadcast stations to get maximum field intensity along the ground, is the best we can do with a single vertical element, giving a useful lobe from about 3 to 27 degrees, ideal for DX work. This height is used in the "Mark II DX Antenna" described later in this book for optimized performance on 14 mHz. A vertical element becomes useless as its height increases above ⅝ wave length, for the low angle lobe shrinks very rapidly, the high angle lobe grows rapidly and splits into several lobes, all of which are at angles too high to be of any value whatever.

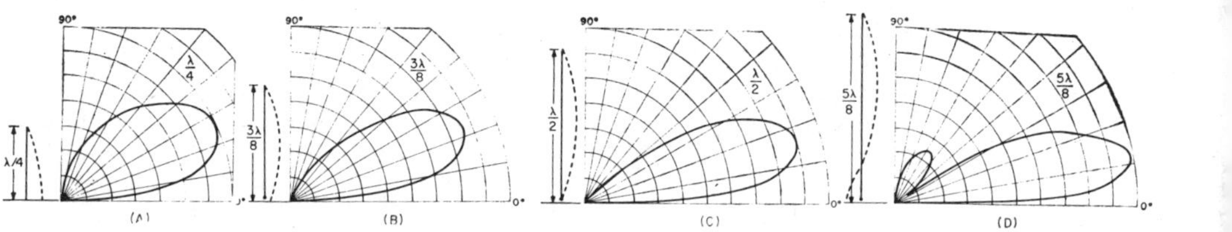

Fig. 4 — Vertical radiation patterns for vertical antennas from quarter to five eighths wavelengths long.

Vertical Stacking

The next logical step is vertical stacking of vertical in-phase ½ wave elements, which results in a colinear antenna. In the early days these were known as Franklin antennas. The free space pattern of a two-element antenna is shown in fig. 5A. It has a gain of 1.9 db over a dipole. The two-element colinear in free space is the same as a ½ wave working against ground (consider the mirror image as the other ½ wave element). There is one factor which is often overlooked when considering vertical antennas working against ground, however. It is the fact that when the mirror image is replaced by a ground plane and all the power is put in the remaining upper half of the antenna, an additional theoretical 3 db gain results. The actual amount of gain thus realized will depend on the conductivity of the ground plane, but with ground of average conductivity we will have some gain, perhaps 1 or 2 db. The three-element colinear has a free space pattern shown in fig. 5B. It has a gain of 3.2 db over a dipole. This antenna in free space is the same as a ½ wave element over and in phase with a ¼ wave element, worked against ground, as shown in fig. 5D. In this latter case our vertical pattern is useful from 3 to 20 degrees. If we go further, and use a four element colinear, or

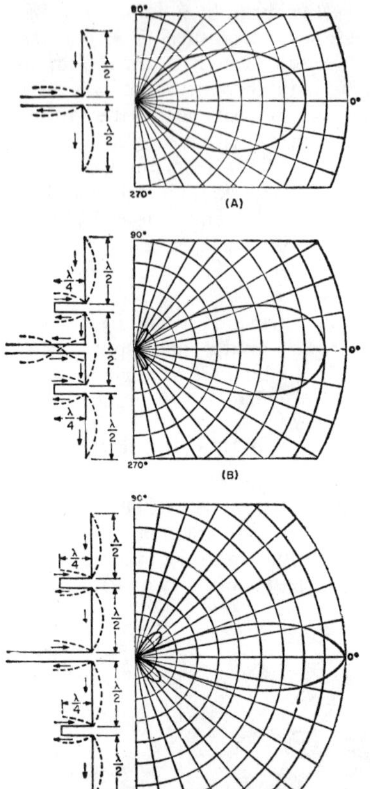

Fig. 5 — Vertical radiation patterns for various types of collinear arrangements. (A) Two elements in free space. (B) Three elements in free space. (C) Four elements in free space. (D) One and one half elements against ground. (E) Two colinear elements against ground.

two ½ wave elements in phase against ground, the patterns of figs. 5C and 5E apply. The vertical angle in fig. 5E is even lower, from 3 to 15 degrees. The gain of a four element colinear is 4.3 db over a dipole.

There are several ways of accomplishing the necessary phase reversal between colinear elements — a quarter-wave stub, coaxial sleeve, or tuned circuit — to make a colinear antenna. To sum all this up, it is apparent that stacking in a vertical plane, of either horizontal or vertical in-phase ½ wave elements, is a way by which one can achieve low angles of radiation in the vertical plane. So, what shall it be — horizontal or vertical? Let's draw some quick but accurate comparisons between the two.

Vertical vs Horizontal

1. Low angles are easily obtainable with a simple vertical antenna.

2. The vertical is simpler in construction. Even 50 or 60 foot self-supporting pipe masts are easily erected.

3. The vertical itself requires less space. Ground radials or a ground plane of some sort are required for efficient operation. However, radials can be bent in directions which fit one's available space.

4. The vertical is easy to feed at its base with unbalanced coaxial feed, using a "gamma" type of feed, or a matching network if required.

5. The vertical discriminates *against* TVI, because TV antennas are horizontally polarized. Some claim that it increases BCI. If this should happen, it is not due to its vertical nature, but to its strong, low angle radiation. *Any* antennas which gives strong low angle radiation, such as stacked Yagis, could also cause BCI.

6. The vertical is somewhat more susceptible to rain and snow static, and to noise impulses in the neighborhood when used for receiving.

7. The vertical is non-directional, and thus cannot discriminate against interference from unwanted directions when receiving. However, one can erect three vertical elements and make a very neat switchable directional array to cover 60 degree sectors in azimuth.

8. The vertical is unobtrusive and pleasing to the eye of one's neighbors.

9. The gain of a colinear vertical can approach that of a three-element horizontal Yagi. It is actually greater at the low angles of interest.

Horizontal vs Vertical

1. Rotatable Yagis can discriminate against interference. Best front-to-back ratio is obtained at the expense of forward gain, however.

2. Relatively large space is required for a 14

mHz beam. On 7 and 3.9 mHz a Yagi is not practical.

3. No ground system is needed for the horizontal antenna.

4. If a rotatable Yagi is used, there is always the problem of rotators freezing or sticking, and of the feedline becoming tangled or twisted.

5. Elements of a Yagi are subject to breakage due to ice or wind loading.

6. The horizontal antenna can give only higher takeoff angles, which are not optimum at the heights used by most amateurs. Lower takeoff angles can be obtained only with vertical stacking of in-phase horizontal elements, which at the heights and spacings required, becomes impractical.

7. For a single horizontal element or even for a Yagi, a height of one wave length is required to get the takeoff angle down to 10 to 20 degrees. Even then, this is not as low an angle as can be obtained with a vertical, and considerable power is wasted in high angle lobes.

8. Horizontal Yagis come in prepared kits, which require little or no mental design effort on the part of the average amateur. This is fine for the non-technical types who do not wish to design or build something which is optimized but which can be made to work after a fashion by almost anyone.

9. A horizontal Yagi on one's house or on a tower in the backyard can be the cause of much unfavorable concern on the part of neighbors. Horizontal antennas can cause much horizontally polarized TVI.

Exploding a Myth

There is another popular myth which I would like to "explode" right now. On the air one will hear fellows say that the signal from a vertical antenna cannot be received on a horizontal antenna as well as can the signal from another horizontal antenna, or vice versa. This statement is parroted as an argument against verticals, for most amateurs use horizontals. Nothing could be further from the truth than this general statement, I assure you. It *is* true that in the strictly line-of-sight case, as in v.h.f. or u.h.f. (in TV reception, for example), polarization of the antennas at each end makes a big difference. However, research has proven that at high frequencies which experience ionospheric reflection in propagating over long distances, the polarization of the transmitted signal, no matter what it originally was, is caused to turn around by the ionosphere in a random manner, and the signal which reaches the distant receiving antenna is no longer linearly polarized, but is elliptically or even sometimes circularly polarized. It has been quite well-proven that on long distance circuits, fading of a signal received on a linear antenna is not so much caused by a change in path loss or attenuation, but it results from a change in polarization of the signal caused by the ionosphere. This opens up the subject of polarization diversity, which might well be covered in detail separately but this is the reason why the Cubical Quad antenna has been used with such success. The Quad consists of orthogonally polarized elements, and it responds well to a signal whose polarization is changing. ■

BASIC PRINCIPLES OF VERTICAL ANTENNAS

Basic principles of vertical antennas are presented in this chapter. Subsequent chapters will cover feed methods, arrays, stacked verticals, broadband types and physical design factors.

IN spite of the fact that there have been many articles written on vertical antennas, including several by myself there are, no doubt, many amateurs to whom their operation is a bit mysterious. There are some amateurs who emphatically and categorically state that verticals are no good without really knowing what they are talking about. But please, I am not trying to sell vertical antennas to the exclusion of other types; I am just trying to spread a bit of knowledge, and we can stand having a bit spread around in amateur radio these days.

For many years I have found the field of antenna design, in conjunction with propagation study, to be a very fascinating one in which to work, both in noval communications and in civilian consulting radio engineering. While on active duty in the Navy Department, Washington, D.C., I had much to do with shipboard antenna problems as head of an office which directed an extensive program of fleet communications improvement. At our service in this endeavor we had the output of years of experience in our naval electronics laboratories and many interesting and valuable contributions from commercial industry. Antenna design is one of the most important factors in the planning of any communications system. In this business one cannot afford to tolerate an "NIH" (Not Invented Here) attitude, nor to say that something is no good. An open mind is the mark of an intelligent man. In these chapters I plan to discuss many aspects of vertical antennas, from some basic theory right on through actual designs, practical configurations, advantages and disadvantages and results to be expected.

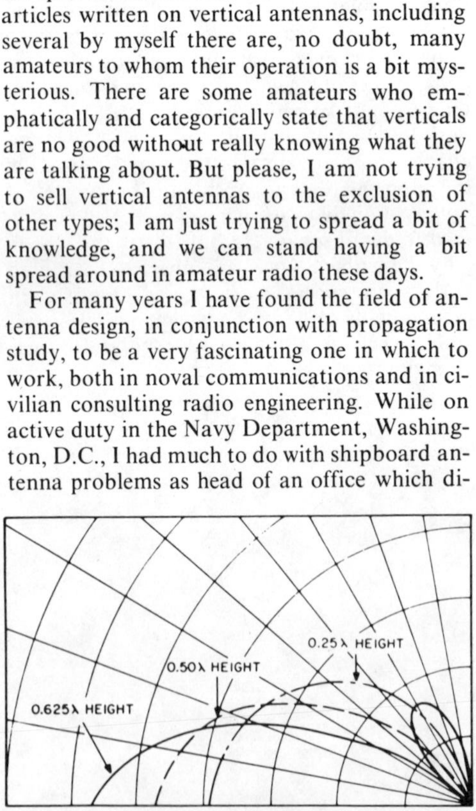

Fig. 6—The theoretical vertical radiation patterns for vertical antennas of ¼, ½ and ⅝ wavelengths in height.

Fundamentally, a vertical antenna is nothing but a horizontal antenna turned up on end. Let us go one step further and rule out the vertical dipole at some distance above earth for practical reasons, and consider the case of the vertical antenna fed at its base against a ground system. For purposes of analysis, this can be considered as a dipole whose lower half has been cut off and replaced by a ground plane. For some purposes the concept of a mirror image (imaginary lower half below ground) is useful. One can say quite truthfully that the total power radiated in free space by a dipole (either vertical or horizontal) has now been concentrated in the upper hemisphere only. Thus it is commonly said that, considering only the upper hemisphere, the radiation from the ground based vertical is 3 db above that from a free space dipole. This is true in one special case, and that is the case of a perfect ground plane with no loss. No such thing exists in actual practice. It can be closely approximated by sea water, or by an extensive copper screen. However, in these two cases there will still be some loss, although minimal. The loss will depend to some extent on frequency. Model studies have shown that for sea water or a large copper ground screen, the vertical radiation pattern of a ground based vertical antenna closely approaches that of the theoretical case.

Vertical Radiation Patterns

What are theoretical vertical patterns of ground based antennas of various heights? They are shown here in figs. 6(A) through 6(C). It will be noted that as antenna height increases, the ground wave or low angle lobe increases until it reaches a maximum at 5/8 wavelengths height. As this height is exceeded, the low angle lobe shrinks very fast, the high angle lobe increases rapidly, and the antenna becomes useless for communications where low angle radiation is required.

Some authors have discussed at length the effect of ground reflections on the vertical pattern, and have stated that all low angle radiation is either absorbed or reflected out of phase, thus making verticals useless for low angle communications. The example invariably used to prove this is one in which the vertical is a dipole, suspended in the air some distance above earth. The center of the dipole is considered to be the theoretical center of radiation. How one is to feed this antenna from a practical standpoint is never mentioned. However this antenna does lend itself very well to a mathematical analysis which can be done by hand or by computer, which shows that ground reflections do occur, and that the vertical pattern has zones of cancellation and reinforcement of radiation, as one would expect. There is an excellent discussion of this in a document[1] which is obtainable from the Superintendent of Documents, Government Printing Office, Washington, D.C. at a cost of $1.25. However, this antenna is hardly a practical configuration. It is an entirely different case from that of the practical, base fed vertical whose ground system is a part of the actual radiating system. For an analysis of this type, consider the theoretical center of radiation to be at earth level, and that the ground system is a part of the antenna system, a lossy part to be

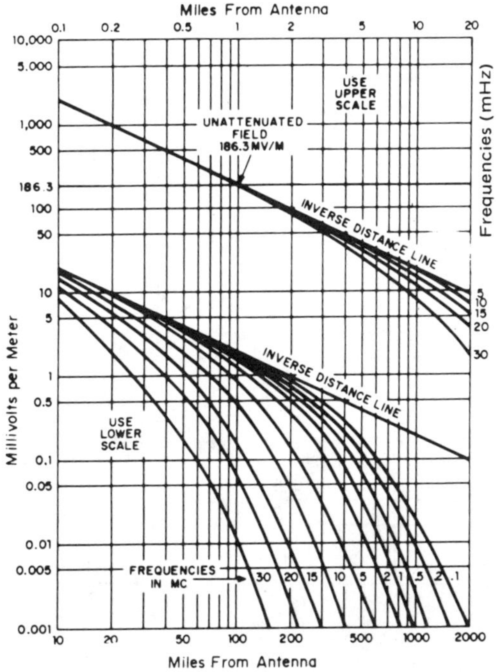

Fig. 7—A plot of ground wave field intensity versus distance from the antenna for various frequencies over sea water based on a radiated power of 1 kw.

sure, but still an inseparable part of it. (The vertical dipole mentioned above has no ground system, nor any connection with earth. That point is always overlooked.)

Ground Attenuation

Ground is lossy. Loss increases with frequency. Ground wave radiation thus decreases with frequency. Figure 7 is a plot of ground wave versus distance for various frequencies, over sea water where ε is 80 and σ is 5 mhos per meter. Figure 8 is a similar plot over "good ground", where ε is 15 and σ is 10 mmhos per meter. Figure 9 is the same plot over "poor ground," where ε is 5 and σ is 1 mmho per meter. (ε is the dielectric constant, and σ is the conductivity.) From these plots it can be seen that at l.f. and m.f. ground losses are fairly reasonable. This is why the FCC considers that the vertical radiation patterns closely approach theoretical at frequencies in the m.f. broadcast band. In past consulting work for broadcast clients before the FCC many skywave patterns have been computed based on this premise, and it might be added that there are literally thousands of actual skywave measurements on file which show this premise

[1] "Siting Criterea For HF Communication Centers," *National Bureau of Standards Technical Note 139.*

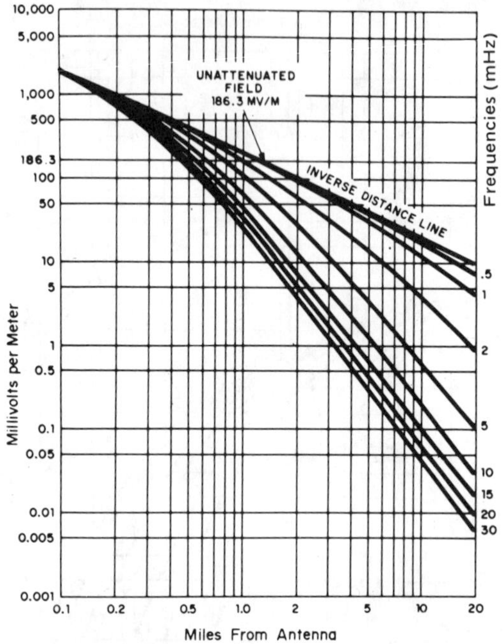

Fig. 8—A plot of ground wave field intensity versus distance from the antenna for various frequencies over "good ground" where $\varepsilon = 15$ and $\sigma = 10$ millimhos per meter, based on a radiated power of 1 kw.

to be realistic. Skywave interference ratios between co-channel stations are computed on this basis. Ground wave coverage and directional patterns are similarly based. More will be said about that later.

It can be seen that at frequencies of interest to amateurs, 2, 4, 7, 14, 21 and 28 mHz, ground wave attenuation over earth increases greatly with frequency. Thus from a ground based antenna, the ground wave will be quite strong within two or three miles, but will rapidly attenuate as distance increases, until it finally becomes useless. There is no such thing at h.f. as 50 to 100 mile ground wave coverage, as occurs in the m.f. broadcast band. Because of these ground losses the vertical pattern from a vertical antenna at h.f. will show considerable "suck-in" at low angles, as shown in fig. 10. The exact patterns will depend on frequency, with more suck-in and loss at higher frequencies.

Thus it is apparent that it behooves the amateur who uses a vertical antenna to install as good a ground system as possible to minimize the losses. He who lives on the ocean beach or other body of water is indeed fortunate, for his local ground losses will be much less than those of the poor soul on a city lot surrounded by buildings, trees, etc.

The vertical radiator with ground system is thus more efficient at the lower amateur frequencies, such as 2, 4 and 7 mHz. It is simply a matter of losses. The vertical radiator with counterpoise or metal ground plane is therefore used at higher frequencies. One does not see v.h.f. or u.h.f. antennas sitting on the ground, with buried ground planes. However, buried ground systems are used up through 28 mHz with good success. More will be said about ground systems and their effect on radiation efficiency later.

Gain

Meanwhile, let's talk about the subject of "gain." What is antenna gain? We hear it mentioned many times on the air, and see it quoted (or misquoted) in advertisements. To have gain, one must start with something and then compare something else to it. What is the start, in antenna work? What is the basis of comparison? What is this "isotropic" antenna we hear about — so many db above isotropic? An isotropic antenna is a theoretical zero length point source. It can never exist. However, it is used as a theoretical and mathematical basis for comparison. In reality, what else could one use? The isotropic antenna is one which radiates equally in all directions. For a given power input, it will produce a certain radiated field intensity at a given distance from it, in all directions. A plot of its radiation pattern is a perfect spherical surface, with the point source at the center of the sphere. Every *practical* antenna, which must have finite length, thickness, and distance from earth, will produce a pattern which is not constant in all directions. Thus the practical antenna is said to possess gain (or loss) in certain directions when compared to the isotropic antenna. *Gain is meaningful only when referred to a specific path or direction.* Directional gain is defined as the ratio of the power that must be radiated by an isotropic antenna to the power being radiated by the test antenna under consideration, to develop equal field intensities from the two antennas in a certain direction. Consider that the isotropic spreads power over the entire spherical surface. The directional antenna, on the other hand, concentrates power in a certain area of the sphere's surface. This concept is shown in fig. 11. Thus for equal field intensities at point "X" within the area illuminated by the beam, the power in the isotropic must be considerably greater than that in the test directional antenna. This power ratio between the two is equal to the *gain* of the directional antenna. It can be seen that there is an inverse relationship between gain and areas illuminated, for a certain fixed power. There is a simple method of computing the gain of a directional antenna.[2] The number of "square degrees" area in the surface of a sphere has been computed to be 41253. Keeping in mind the relationships mentioned above, divide this number by the number of "square degrees" in the beam of the directional antenna between the half-power points, and the result will be a close approximation of the power gain of the directional antenna, in the direction of the beam. Consider the 3 element horizontal Yagi antenna, whose beam width is about 60 degrees in the horizontal plane,[3] and about 30 degrees in the vertical plane, for a height above earth of ½ wavelength. The approximate area within the half-power limits is 1800 square degrees.

$$\text{Power gain} = \frac{41253}{1800} = 22.9,$$

or 13.60 db above isotropic

Note that I said *above isotropic,* the theoretical antenna which cannot exist. Sometimes, without much thought, the horizontal dipole is used as a reference. But this can be very confusing, because its radiation varies with direction, both in the horizontal and the vertical planes, and with distance above earth. Thus, a manufacturer can easily confuse an amateur by saying simply that his beam has so much gain in db. Gain with reference to what? To isotropic? To a dipole? In what direction? At what vertical angle? At what height above earth? The "gain above isotropic" figure is always more impressive to the reader, because in the direction of its main lobe the dipole has several db gain with reference to isotropic. Thus gain figures quoted by manufacturers have to be considered with care, with special attention to the fine print. Don't be fooled by gain figures; ask the question, "Gain with reference to what?" You may be surprised at the answer. Some beams, especially miniature ones, have little gain above a dipole, possessing only *directional discrimination.*

We might be inquisitive, not to prove anything but to simply find the answer, and compute the gain of a ⅝ wave vertical antenna. We must decide on a reference, which by rights should be isotropic. The vertical beam-width

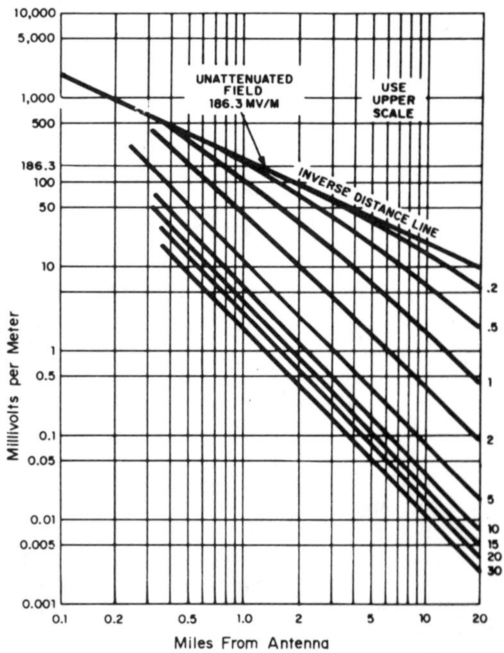

Fig. 9—A plot of ground wave field intensity versus distance from the antenna for various frequencies over "poor ground" where $\varepsilon = 5$ and $\sigma = 1$ millimhos per meter, based on a radiated power of 1 kw.

to be expected is about 15 degrees. Since the horizontal spread is a full 360 degrees, the approximate number of square degrees is 5400.

$$\text{Power gain} = \frac{41253}{5400} = 7.63$$

or 8.84 db above isotropic

Note that I again said *above isotropic,* the theorectical antenna which cannot exist. Thus by computing the gains of individual antennas with reference to isotropic, their gains relative to each other may be determined. The above computations show a relative gain of 4.76 db for the Yagi over the ⅝ wave vertical, but *in different main lobe directions.* The lobe from the ⅝ wave vertical is lower than that from the Yagi.

Gain is meaningful only when considered with reference to a particular communications path and set of propagation conditions. This factor can never be overlooked. A simple illustration will suffice. Antennas designed for the tropical broadcasting bands to produce sky-wave coverage over a certain Latin American country or two usually are horizontal dipoles at a low height above ground, thus producing

[2] Terman and Pettit, *"Electronic Measurements,"* Second Edition p. 436.
[3] Kraus, J. D., "Antennas," *McGraw-Hill,* p. 321.

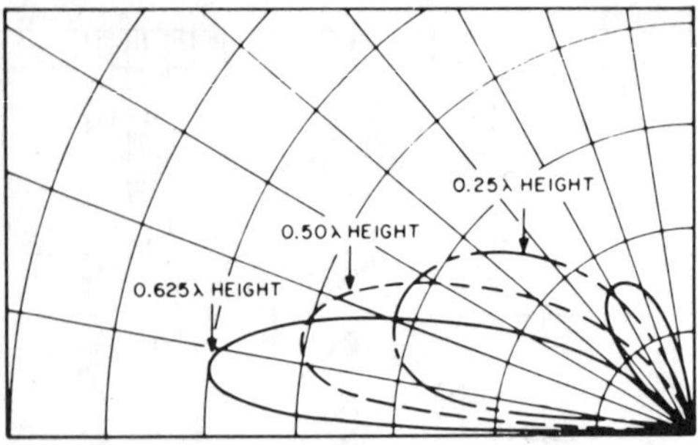

Fig. 10—Ground losses cause "suck in" of the vertical radiation pattern of the vertical antenna. The effects are shown above for antennas of various wavelengths.

only high angle skywave lobes. They are useless for long distance communications. They are designed that way intentionally, to restrict coverage. Thus, one cannot say that one antenna is no good, or that another is the perfect answer. Each type has its own specific application, designed to fit certain systems requirements.

There is another way of determining the efficiency of a vertical antenna. As a basis of evaluation in this case, it has become standard practice throughout the years to use the "unattenuated field intensity at one mile" as the reference for field intensity measurements. This figure is the basis for all plots such as those in figs. 7, 8, 9 and 12. It is a figure of merit which can be accurately determined by a carefully made series of field intensity measurements, taken along a number of radials (usually at least 8) and out to considerable distance from the antenna. Each set of radial measurements is then plotted in a field intensity plot such as that in fig. 12. From knowledge of the ground conductivity in the area (shown in fig.

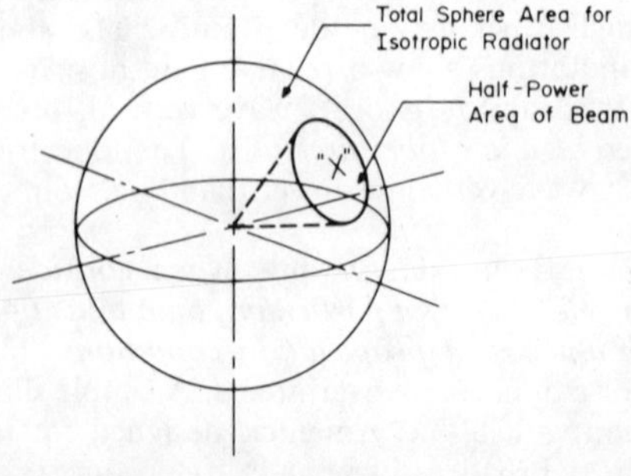

Fig. 11—Comparison of radiation pattern of an isotropic versus a directional antenna.

16) the plotted curve can be fitted against one of those in fig. 12, and the actual unattenuated field being radiated can be determined. The curves of fig. 12 show field intensity versus distance for various ground conductivities, with the *standard unattenuated field of 186.3 mv/m at a mile for one kilowatt radiated power.* Figure 12 is an adaptation of one of a series of such figures from the FCC rules for various frequencies in the m.f. broadcast band. It is used merely as a convenient example. Similar curves can be plotted for frequencies in the h.f. range.[4] As may be seen, figs. 7, 8 and 9 are a partial plot of such. There is a direct relationship between them and fig. 12.

One may ask the origin of the "186.3 mv/m at one mile" for a theoretical zero-height point source current element on a ground plane. This figure is the unattenuated field at one mile for a theoretical zero-height point source current element on a ground plane. The derivation of this figure is given in one of the references.[5] The table below shows the relationship between the zero height point source current element and the isotropic. The current element is, of course, something which cannot exist in practice. Practical antennas have finite

Type	Pattern	mv/m at 1Mile (1 kw)	Power Gain	DB Gain
Isotropic	⊙	107.6	1	0
Half Isotropic	(symbol)	152.1	2	3.01
Current Element	(symbol)	186.3	3	4.77

height. Figure 13 shows the vertical patterns which can be computed for antennas of various heights. It may be seen that this figure is the same as fig. 6, but plotted with field intensity values. The theoretical unattenuated field at one mile along the horizontal plane may be read for each. They are as follows:

Height of antenna	*Unattenuated field at One Mile (1 kw)*
0.25 (¼)λ	196 mv/m
0.50 (½)λ	236 mv/m
0.625(⅝)λ	276 mv/m

[4] Norton, K. A., "Calculation of Ground Wave Field Intensity Over a Finitely Conducting Spherical Earth," *Proc. of I.R.E.,* Dec. 1941, p. 632.

[5] Brown, G. H., et al, "Ground Systems As A Factor In Antenna Efficiency," *Proc. of I.R.E.,* June 1957, p. 735.

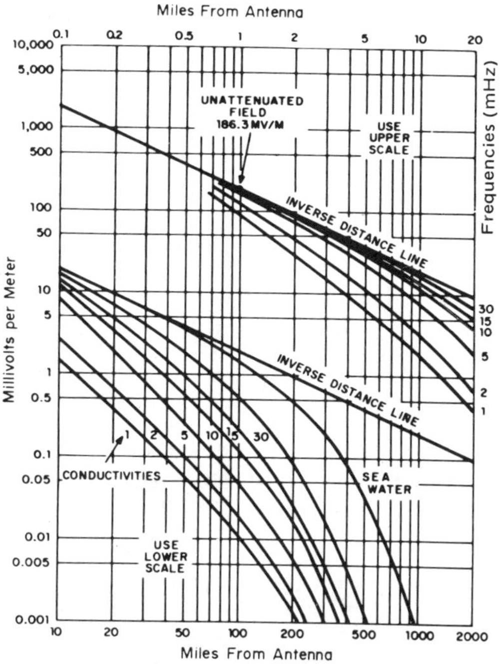

Fig. 12—A plot of ground wave field intensity versus distance from the antenna for 970 to 1030 kc. The unattenuated field is 186.3 mv/m at one mile for a radiated power of 1 kw. The dielectric constant ε is 15 and the conductivities σ are shown in millimhos/meter.

Thus it may be seen that the 0.625 (⅝) wave antenna produces the greatest ground wave field, and the lowest main lobe. Figure 13 is taken directly from the FCC rules. The values shown are theoretical maximums. Antennas with adequate ground planes and low losses will closely approach these values, as will be seen shortly.

The Ground System

This leads to the next item of my discussion, which is the ground system. There have been several good articles published in past years on the subject of ground systems and their effect on antenna efficiency and other parameters. One of the best, which we in the business consider to be the "bible," is by G. H. Brown.[5] It was based on an extensive experimental program of actual field measurements by RCA Laboratories. It covers such factors as numbers and lengths of radials, depth of burial, ground currents, losses, radiation resistance, and other factors. The development of the figure of 186.3 mv/m is shown. Figure 14 , based on this reference, shows a factor of interest to

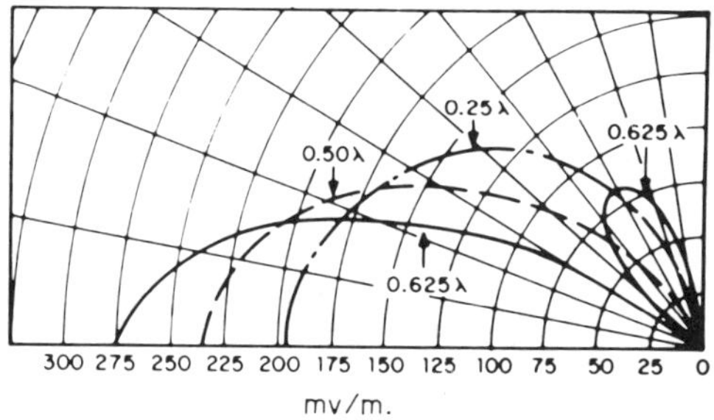

Fig. 13—Theoretical vertical patterns for antennas of 0.25, 0.5 and 0.625 λ heights.

amateurs, which is the effect of changing the number of radials of a fixed length. Figure 15 shows another such factor, the effect of varying the length of the radials, for certain fixed numbers of radials (15 and 113). A study of these figures shows that radiation efficiency is greatest for a large number of long radials, but it can be seen that the curves flatten off and that there is an area of diminishing returns where the percent increase in efficiency is not really worth the added cost of copper in the ground. FCC rules require that broadcasters use at least 120 radials ¼ wave long. Many stations better this. For the average amateur, 16 is a reasonable and economical number. Doub-

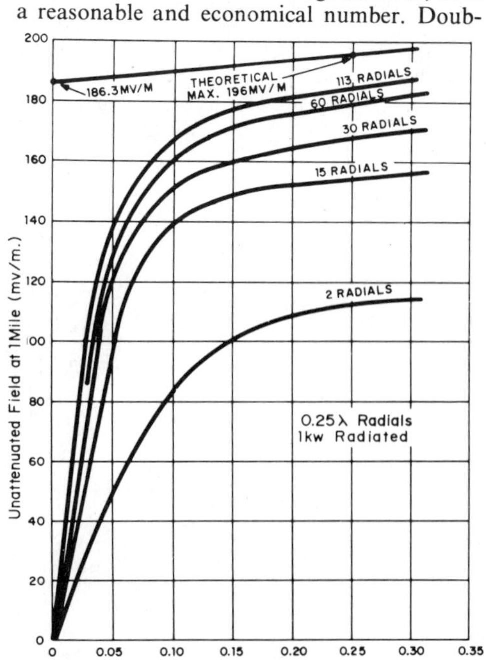

Fig. 14—Effect on the unattenuated field at one mile by changing the number of radials.

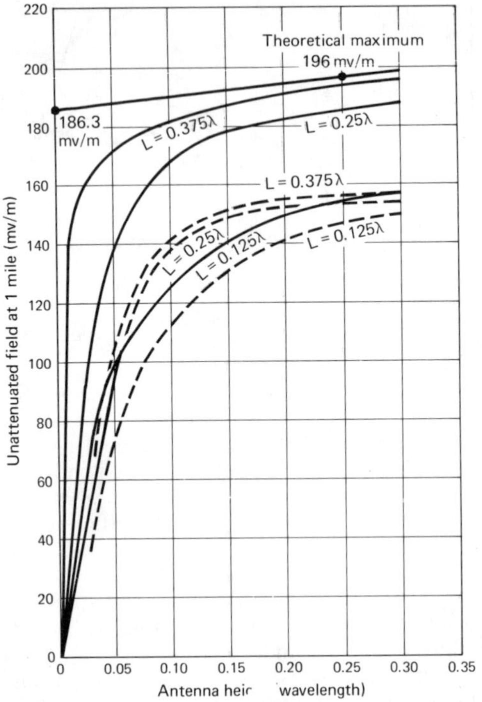

Fig. 15—Effect upon the unattenuated field strength at one mile by varying the length of the radials. The radial lengths, L, are specified in wavelengths. Solid curves are for 113 radials and the broken curves are for 15 radials.

ling the number will not double the radiation efficiency.

It should next be realized that the unattenuated field determined by measurements is the "starting out" radiation from an antenna of given height. It is less than the theoretical field by an amount which is determined by the losses in the ground system, as shown in figs. 14 and 15. It is also affected by losses in the antenna itself, and by the presence of buildings, wires, and other absorptive elements in the vicinity of the antenna. This is especially true at h.f., where losses in the immediate vicinity may be quite high, as some of those objects approach an appreciable size in terms of wavelength. He who has a good, clear site is fortunate. Then, as the signal proceeds further from the antenna along the surface of the earth, it is attenuated even more by ground losses, causing the considerable suck-in of the low angle radiation, especially at the higher frequencies.

Figure 16 is included for those who are interested in knowing something about ground conductivities in the United States. It is taken directly from the FCC rules. This map is based upon thousands of field intensity measurements made and on file with the FCC. It will be noted that some of the midwest states have ground conductivity as high as 30 mmhos/-meter. On the other hand, mountainous areas are quite low in conductivity, around 2 or 4 mmhos/meter. ■

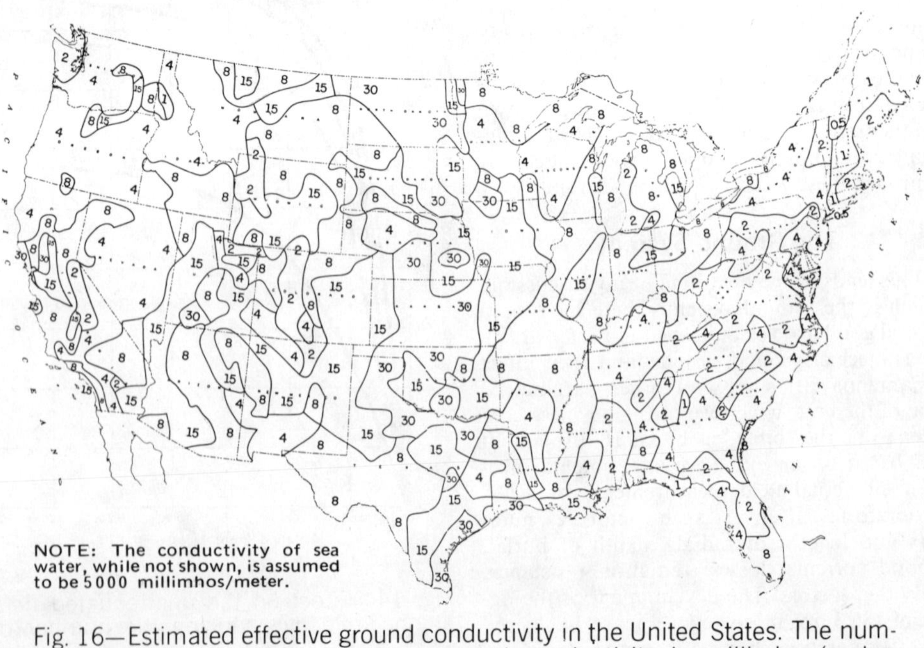

Fig. 16—Estimated effective ground conductivity in the United States. The numbers represent estimated effective ground conductivity in millimhos/meter.

BASE IMPEDANCES AND METHODS OF FEEDING AND MATCHING

Preceding chapters covered basic principles of vertical antennas. This one deals with base impedances and methods of feeding and matching.

THE input (base) impedance of an ungrounded vertical antenna which is being fed against a ground plane is dependent on several factors. The one factor which has the greatest effect is of course the height of the vertical radiator. The height is usually mentioned in terms of wavelengths. In the second chapter vertical radiators of 0.25, 0.50 and 0.625 wavelengths height were discussed from the standpoint of their radiation characteristics. It is of interest, however, to plot the base impedance of vertical radiators of all heights from 0.10 through about 0.65 wavelengths height so that the variations may be noted in detail.

The second factor which affects base impedance is that of height (length) versus thickness (diameter). It is customary in texts to define this in terms of the ratio *length/diameter (L/D)*. If one considers the vertical radiator being fed against a ground plane as an open-ended, unbalanced, lossy transmission line, which it actually is, it will be realized that it must have a characteristic impedance like any other transmission line. The magnitude of this characteristic impedance is determined by the conductor size and spacing. In the case of the vertical antenna (considered as a transmission line) the spacing (position of the vertical conductor relative to the ground plane) is fixed. The only factor which can then be varied, to have an effect on the characteristic impedance, is the vertical conductor's diameter. The effect on the characteristic impedance of changing diameter can be computed. Seveal authors treat this, but perhaps Schelkunoff[6] is one of the best references.

Radiation Resistance

If this unbalanced transmission line could be terminated in its characteristic impedance at its far end, one would see the characteristic impedance when measuring at its input connection. However, it cannot be terminated. It must remain open-ended. Thus it has standing waves on it, and it radiates. When speaking of it as "lossy", this does not refer to its inherent I^2R losses. This refers to the fact that power put into it is "lost" or radiated into space. This power is represented theoretically as the power dissipated in a *radiation resistance*, R_1. For a good discussion of the concept of radiation resistance, see Laport.[7] The radiation resist-

[6] Schelkunoff and Friis, *"Antenna Theory and Practice,"* John Wiley and Sons.

[7] Laport, E. A., *"Radio Antenna Engineering,"* McGraw Hill.

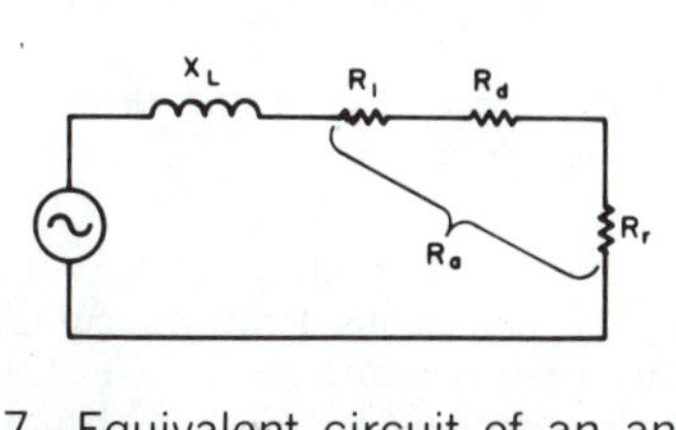

Fig. 17—Equivalent circuit of an antenna system. The total antenna resistance, R_a, is equal to the sum of R_1, R_d and R_r.

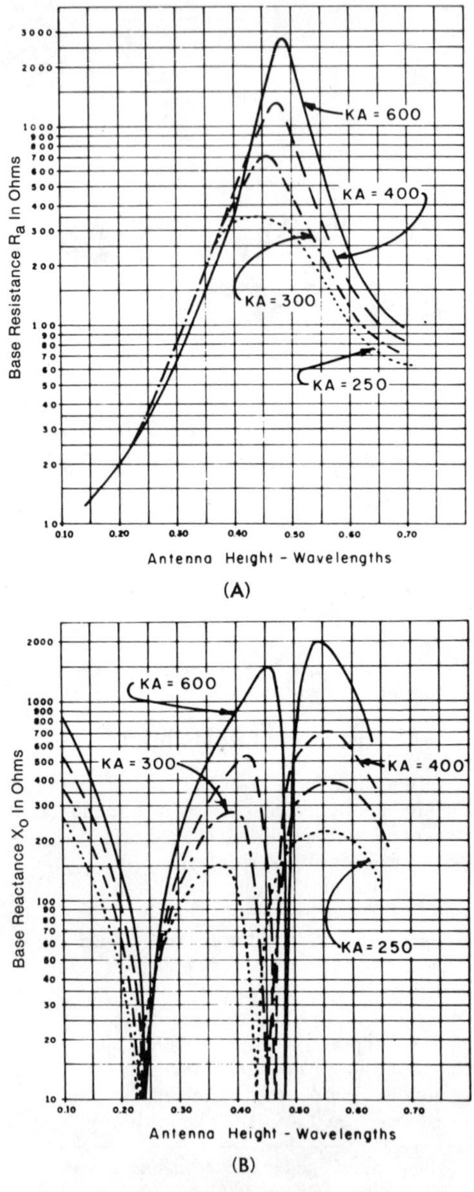

Fig. 18(A)—Vertical antenna base resistance versus antenna height in wavelengths for different values of characteristic impedance, K_a. (B)—The plot of the base reactance versus antenna height for the same values of K_a

ance of an antenna cannot be measured. It can be computed with considerably accuracy. The radiation resistance times the square of the current at the point of reference is the radiated power. This radiated power can be determined quite closely by making field intensity measurements along radials at varying distances out from the vertical radiator, plotting them on field intensity paper, and determining the unattenuated field intensity at one mile, as discussed earlier.

When measuring into the antenna, one sees the total antenna resistance, R_a, which is composed of several parts. This is depicted in fig. 17, which shows the equivalent series circuit of an antenna possessing some inductive reactance. These parts are the radiation resistance R_r, the ohmic loss resistance R_l and the dielectric loss R_d. In the m.f. and h.f. cases, the R_r is quite large, and it makes up most of R_a. In fact, in a well designed installation, the several loss resistances are usually neglected. However, at l.f. and v.l.f., where the vertical radiator is a small part of a wavelength, R_r is very small. In fact, at v.l.f. it is often as small as a fraction of an ohm. When this is combined with fractional ohm loss resistances, efficiency is reduced considerably. At v.l.f. it requires very special care in design, plus many model studies, to obtain an efficiency as high as 50% or so. Not so at h.f., however. We are up in the range of tens and hundreds of ohms for R_r.

The standing wave configuration on a vertical radiator will depend on its height. The base input impedance will, therefore, show variations with respect to frequency. The magnitude of the peak excursions of impedance will depend on the radiator's characteristic impedance. This fact is shown in fig. 18(A) and 18(B), which are plots of R_a and X_a, respectively. They are plotted separately for clarity. The family of curves shows the effect of varying the characteristic impedance of the radiator. Characteristic impedance is denoted by the factor K_a which can be computed from the following equation:

$$K_a = 120(\log_e \frac{21}{a} - 1)$$

Values of K_a have been computed as follows:[6]

L/D	10	50	100	200	300	600	1000
K_a	323	516	599	682	731	814	875

By reference to fig. 18(A) and 18(B) it can be seen that a thin vertical radiator exhibits greater variations of impedance than does a thick one. (This is true of any antenna.) We make use of this fact in designing broadband antennas, which will be covered later in this book.

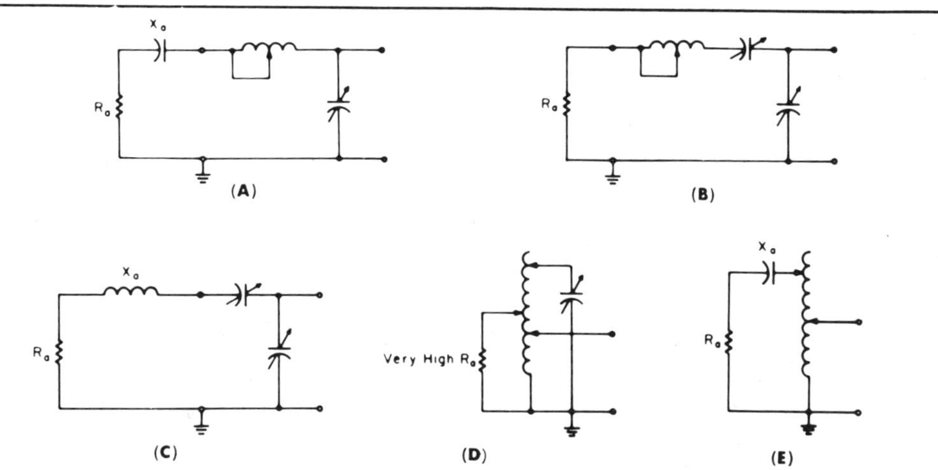

Fig. 19(A)—Matching unit for an antenna with R_a -j X_a = R (low) -jX_c.(B)—Matching unit for a $\lambda/4$ resonant antenna with a low R_a only. (C)—Matching unit for an antenna with R_a + JX_l. (D)—Matching unit for an antenna with a very high R_a, $\lambda/2$ resonance. It is difficult to match well and should be avoided if possible. (E)—Matching unit for an antenna with a high R-jX_c(higher than coaxial line impedance).

Bandwidth

This brings up the subject of antenna bandwidth. The bandwidth of an antenna is dependent upon the rate of change of reactance versus the rate of change of resistance, with frequency. It may be seen that the thick antenna, the one with the lower K_a, will have a greater inherent bandwidth than the thin one. For practical purposes, operating bandwidth is the band of frequencies centered about the tuned resonant frequency, enclosed within certain v.s.w.r. limits which depend on other factors such as transmission line capabilities, transmitter v.s.w.r. tolerance, modulation r.f. passband response requirements, *etc.* A figure which is commonly used is the bandwidth to the "3 db down" power limits. There is a very good discussion of antenna bandwidths in Laport.[7]

In the l.f. and v.l.f. cases, bandwidth restrictions can seriously limit an antenna's transmission capabilities unless special designs are adopted to increase bandwidth. These restrictions are caused by the high Q of these antennas. With very high capacitive reactance and very low resistance, the l.f. or v.l.f. antenna is a high Q device. There are special designs, which will be covered later, which improve the bandwidth of these antennas, but at the expense of efficiency. One may ask why we need greater bandwidths at l.f. and v.l.f. than heretofore. The reason is that we are gradually converting to multi-channel teletype transmission systems at these frequencies, to increase our traffic handling capabilities, and such systems require greater bandwidth than the old method of single channel keyed c.w.

At m.f. and h.f. we are not worried too much about bandwidth, except that in the case of some m.f. directional arrays; the coupling and branching networks may in some cases restrict the antenna system's bandwidth, and cause loss of sideband modulation frequency response, unless special design precautions are taken. It is for this reason that the FCC requires the consulting engineer to measure and plot R_{in} and X_{in} for a directional antenna system, versus frequency. At h.f. there is no such problem. In the h.f. case bandwidth is important from the standpoint of avoiding retuning of the antenna when shifting frequency, which is very desirable in military and commercial installations.

Reference should again be made to figs. 18(A) and 18(B). Now that we have the plots of impedance versus frequency (plotted in terms

Height	*Base Impedance*
Less than 0.25λ resonance.	*R* — *jXc*
0.25λ resonance.	*R* + *j0*
0.25λ physical length.	*R* + *jXL*
0.25λ resonance to 0.50λ resonance.	*R* + *jXL*
0.50λ resonance.	*R (Very high) (Not recommended)*
0.50λ physical length.	*R* — *jXc (High)*
0.50λ physical length to 0.625λ electrical length.	*R* — *jXc*

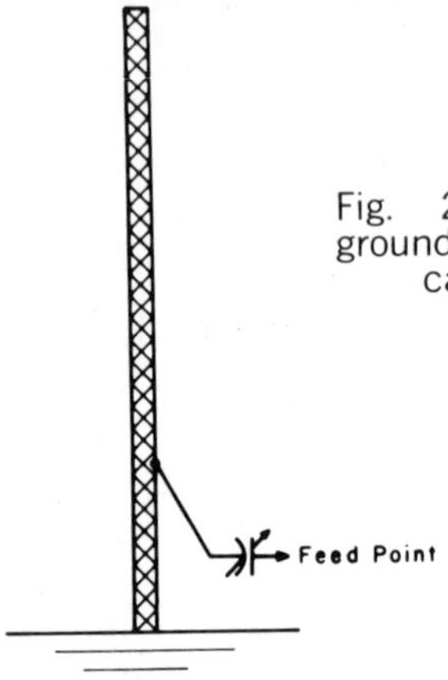

Fig. 20—Shunt fed grounded base vertical antenna.

of height of radiator in wavelengths), we can state the following general rules for the kind of base impedance which will exist: (see chart on bottom of page 25.)

It will be noted that the resonant length is always less than the physical length by an amount which is inversely dependent on the *L*/*D* ratio.

Tuning And Matching Networks

This leads to the next topic, which is the tuning and matching networks required to resonate and match the antennas of heights shown in the preceding impedance summary, to a known source. In modern times the known source has come to be the nominal 50 ohm coaxial line, although 70 ohm line is sometimes used. The matching networks are quite simple, and are shown in figs. 19(A) through 19(E). The values of the reactances required in each case may be calculated by use of r.f. network theory as shown in Terman.[8] In general, if the antenna input reactance is $-jX_c$, inductive reactance $+jX_l$ will be required in the network to tune it, and *vice versa.* The network also simultaneously transforms the antenna input resistance into a value which matches the characteristic impedance of the transmission line.

Where an antenna is fed through a tuning and matching unit of this type, the *total antenna system reactance* must be taken into account in bandwidth computations. This total value includes not only the antenna's own input reactance versus frequency, but also the effect of the reactances in the matching unit as they vary with frequency. The computation of this is not difficult, although perhaps a bit laborious. The simplest approach is to actually tune up the matching unit with antenna attached, and then measure the matching unit's input impedance versus frequency. There are certain "tricks" with networks which can be employed to improve antenna system bandwidth and give lower v.s.w.r. over a range of frequencies.

Shunt Fed Verticals

In the previous paragraph we spoke of tuning and matching a series fed, ungrounded, vertical radiator. Now let's consider the case of the grounded base, shunt fed vertical radiator. There have been several excellent articles on this subject.[9,10] The shunt fed configuration is shown in fig. 20. Simply stated, a tap is brought off the vertical radiator at a distance above ground and it is used as a feed point, having its input impedance matched to the characteristic impedance of the transmission line. In some cases it is possible by trial and error to adjust the tap's vertical position on the radiator so that the input impedance is $50 + jX_l$, thus permitting an easy match to 50 ohm line by use of only a series capacitor of equal reactance. This adjustment, however, can be quite tedious and critical, especially when hanging on the side of a tower or mast, and it is much simpler to use a matching network of proper reactance values. This type of feed has been used in many broadcast installations. It has been proven by actual full scale model studies and field intensity measurements[9] that use of this feed method is just as efficient as is series feed, and that the horizontal radiation pattern is not distorted by the presence of the sloping feed wire. By both mathematical analysis and actual radiator current measurements[9,10] it has been shown that current distribution on the shunt fed radiator closely approaches that of a series fed radiator, and that the vertical radiation pattern is not appreciably different.

A similar arrangement, and one most often used by amateurs, is the "gamma" match or "omega" match, similar to that used on the driven element of Yagi arrays. This is shown in fig. 21. This principle was used in the feed arrangements for the Mark III and the Mark IV DX antennas which will be described later in this book. A feed rod is connected to the radi-

[8] Terman, F. E., *Radio Engineering Handbook,"* McGraw Hill (first edition).

[9] Morrison, J. F., Smith, P. H., "The Shunt Excited Antenna," *Proc. of the I.R.E.,* June 1937.

[10] Badoux, P., "Current Distribution and Radiation Properties of a Shunt Excited Antenna," *Proc of the I.R.E.,* June 1940.

ator at a suitable point some distance above ground, and it is brought down the side of the radiator rigidly supported on standoff insulators. It is fed at its bottom end. It must always be considerably less than ¼ wavelength for its input impedance to be of a reasonable value. The input impedance will always measure R + jX_l, because the short section of transmission line (feed rod plus the portion of radiator concerned) thus formed is less than ¼ wavelength and shorted at its top end. Matching then becomes a question of adjusting the several capacitive reactances in the matching unit. The *R* term of the input impedance may be quite low.

If the feed rod (or wire) runs all the way to the top of the vertical radiator, the whole thing becomes a folded unipole antenna and was used as a feed in the Mark IV for 2 and 4 mHz. This type will be discussed in detail in a later chapter.

Ground Bus

There is one additional point that should be made very emphatically in connection with the shunt fed and gamma or omega fed antennas. That is that the ground bus from the tuning unit to the base of the radiator should be heavy and low loss. This bus carries considerable current, especially where the *R* term of the feed point impedance is low. For 1 kw in the antenna (2 kw p.e.p. transmitter input), with a feed point *R* of 10 ohms, there would be 10 amperes current flowing in this bus (and in the feed rod as well). The capacitors in the matching unit should be large enough to withstand the voltages developed across them by this high current.

Each method of feed has certain advantages and disadvantages. The series feed has the advantage of simple matching when the antenna base impedance is capacitive, by means of a single tapped coil, as in fig. 19(E). A simple variable matching unit with a band-switching arrangement can be built to tune a series fed radiator at several frequencies. However, unless a 3-legged self-supporting tower is used, the series fed radiator must usually be guyed because it has to be insulated from ground. Of course one may use small self-supporting

Fig. 21—Omega fed grounded base vertical antenna.

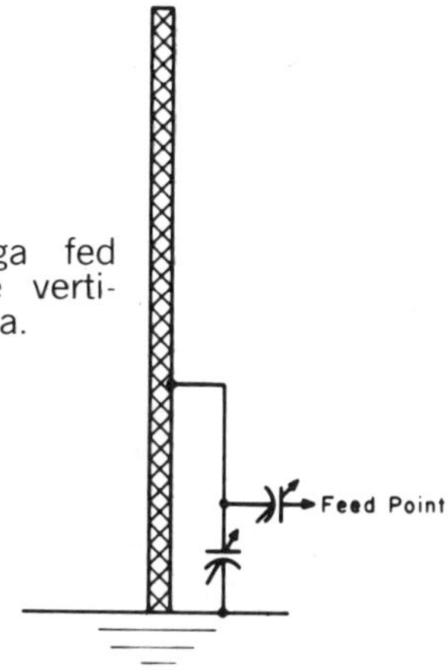

whips for short verticals.

On the other hand, the grounded base radiator is quite easy to make self-supporting. Short, light weight ones can be set right in the ground itself, and tall heavy ones can be set in concrete. Very tall ones must usually be guyed. The grounded base permits other circuits to be run up the tower or mast without any special choke coils or other isolation devices. Such circuits would be tower lights, rotary beam motor control cables, and in some cases even the coaxial lines for v.h.f. or u.h.f. antennas can be run up a grounded vertical radiator without isolation. This latter idea has been used in several broadcast tower installations where a.m., f.m. and c.a.t.v. antennas utilize the same tower structure.[11]

Feed rod construction is simple. Even multiple feed rods are not difficult to install and adjust for feeding the vertical radiator on several frequency bands. Jasek's "Antenna Engineering Handbook"[12] is a veritable gold mine of information on vertical radiators (and on other types also), and especially on some of the schemes we have been discussing. It also provides an excellent source of additional references, listed at the end of each chapter. ■

[11]Gureckis, P. V., "A Simple Method of Isolating CATV Antennas On Standard Broadcast Towers," *Broadcast Management-Engineering,* January 1968.

[12]Jasik, H., *"Antenna Engineering Handbook,"* McGraw Hill.

4

SHORT VERTICAL ANTENNA CONSIDERATIONS

This chapter covers short vertical antennas, their limitations and ways to overcome these limitations. Several practical designs and methods of feeding are shown.

SHORT vertical antennas are of interest to many amateurs. Every mobile whip antenna is short, much too short, at frequencies in the 7, 4 and 2 mHz bands. Also, not everyone is blessed with sufficient clear space for erection of horizontal dipoles at those frequencies. For such amateurs, the short vertical is a solution which may be attractive, especially for use on 4 or 2 mHz. This discussion is for those who care to know something about how such antennas can and do work, and what their limitations are.

Since this type of antenna is electrically short, something has to be done to lengthen it so that it will resonate and take power from the transmitter. One of the simple ways of resonating a short antenna is by use of a series loading coil at the base of the antenna. Why is a coil used? Reference to figs. 18(A) and 18(B) will show that the input impedance of an antenna shorter than ¼ wavelength consists of a fairly low resistance and a capacitive reactance. The shorter the antenna, the lower the resistance and the higher the reactance. Thus, the shorter the antenna, the more loading coil is required, and, since loading coils have inherent loss resistance, the lower will be the efficiency of the antenna system. Remember these facts, for we shall talk about them later on.

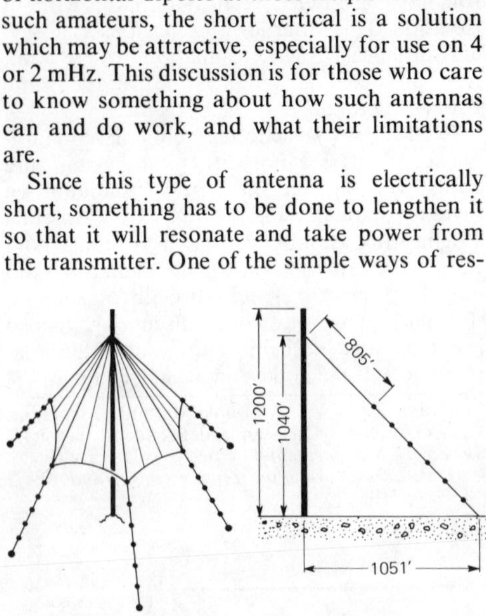

Fig. 22 — Top loading a vertical by the use of guys. The rear guys, also broken by insulators, are not shown for the sake of clarity.

One of the common forms of short antennas is the mobile whip. Early mobile whips were built with base loading coils, and for changing bands it was possible to change loading coils by unscrewing the whip and the coil from the spring mount. These worked after a fashion, but were quite inefficient at 7 and 4 mHz. They were inefficient because the current distribution on the vertical whip was poor. With the antenna in resonance the point of maximum current, or "current loop", was at the bottom of the loading coil. In any antenna, the portion which carries the greatest current does most of the radiating. Thus with the base loaded whip, the greatest current was in the loading coil, merely creating an inductive field, while a relatively small current was flowing in the whip itself. The answer to this dilemma was to make the whip in two parts, and put the loading coil up in the whip itself, thus giving at least the bottom portion of the whip a chance to carry some appreciable antenna current and do some radiating. Most mobile whips today are made in this fashion. You can see them in the ads in this and other magazines, and in distributor catalogs. In spite of all the claims made by the manufacturers, however, even this type of whip which is so very short electrically is quite inefficient on 7, 4 or 2 mHz. Factors which contribute to the inefficiency are lack of adequate ground plane, losses in the loading coil, and the short length of the portion of the radiator carrying greatest current.

Top-Loading

A better approach to the problem of increasing the radiation from a short antenna is by use of top loading. In some cases of m.f. and l.f. high power installations, an actual top hat has been constructed on a tower by using steel structural members, supporting a horizontal web of connecting wires. A much more practical approach is the use of the upper portions of top level guys as top loading by connecting them to the tower and breaking them with insulators a certain distance down from the top. There have been many studies made in an effort to determine optimum configurations of top loading cable length, angle of top loading cables from the tower, number of top loading cables, and interconnection of the lower ends of the top loading cables. One of the most extensive of these was done by Smith and Johnson.[13] That article contains many graphs of resistance and reactance for various top loading guy configuration, plus discussions of the effects of insulator losses and ground system losses. The results of this study showed that substantial gain in radiated power can be realized by this "umbrella" type of top loading. Connection of a wire skirt around the lower ends of the top loading cables shortens the length of cables required to produce a particular result. In cases of high power at l.f., use of the skirt wire reduces corona losses. Top loading gets the current up in the antenna, where it belongs.

Another study which showed substantially the same thing was one made on a 1200 foot tower at Forestport, N.Y.[14] Optimum configuration of top loading cables for maximum resistance and lowest reactance (best efficiency and bandwidth) is shown in fig. 22. The tower is shown being guyed in six directions at the

[13]Smith, C. E., Johnson, E. M., "Performance of Short Antennas," *Proc of IRE*, Oct. 1947.

[14]"Forestport Antenna Study," Technical Report 42-F Contract AF 30 (602)-2588, 31 Jan. 1962. Available to defense contractors through D.D.C. (Unclassified)

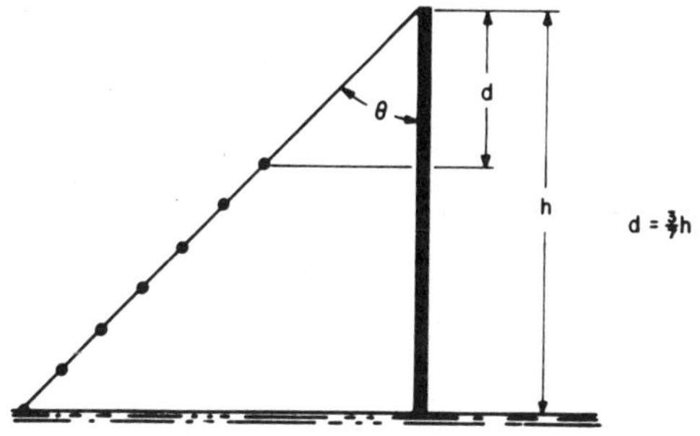

Fig. 23 — Critical dimensions for the top loaded antenna.

1040 foot level, with the guys at a 45 degree angle with the tower. A skirt cable is used, and between each pair of guys there are two additional top loading cables, thus making a total of eighteen top loading cables. The top loading portion of each guy is 805 feet long. The study shows the 1200 foot tower to be resonant (zero reactance) at 80 kHz with this top loading configuration. Now, scaling the dimensions down for use on 2 mHz, the tower height becomes 48 feet, and the top loading guy cable length becomes 32 feet. This is a reasonable size for amateur use on 2 mHz. For 4 mHz these dimensions would be halved. The importance of an adequate ground system under such a radiator cannot be overemphasized. As shown in Chapter Two the number and length of radials has great effect on radiation efficiency, especially in the case of short vertical radiators.

Belrose conducted a study of top loading parameters [15] and determined that radiation resistance was maximum when $d = 3/7h$ and θ

[15]Belrose, *et al*, "Engineering of Communications Systems for Low Radio Frequencies," *Proc of IRE*, May 1959.

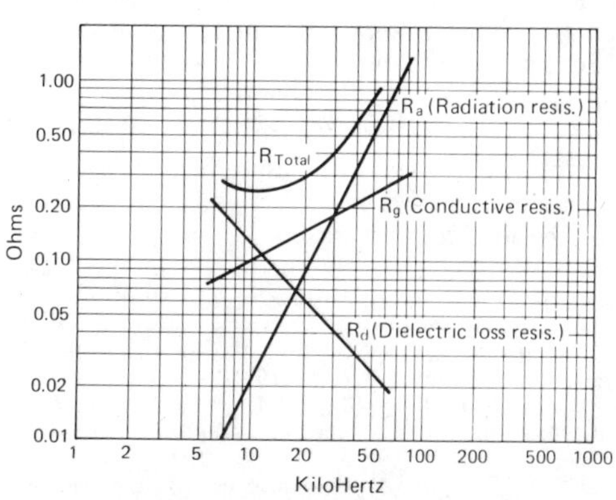

Fig. 24 — Resistive Components of a v.l.f. antenna.

Type A = 3 guys
Type B = 6 guys
Type C = 6 guys with skirt wire

Type	H	L_1	L_2	S	d_1	d_2
A	.124λ	.122λ	.172λ	.013λ	.0012λ	.000187λ
B	.104λ	.102λ	.144λ	.011λ	.0010λ	.000156λ
C	.081λ	.079λ	.112λ	.0036λ	.00078λ	.000121λ

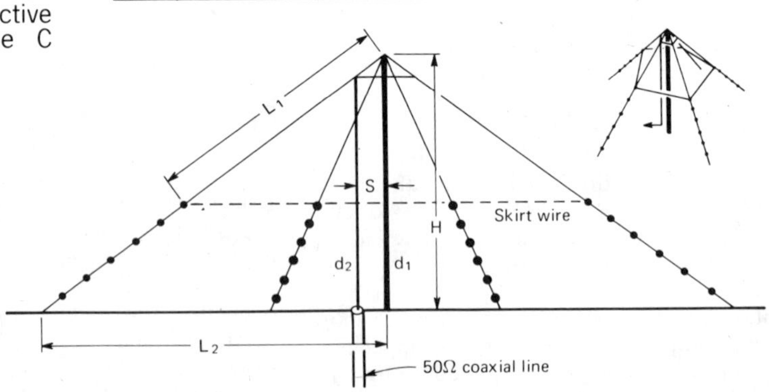

Fig. 25 — Top loaded Folded Unipole developed at the N.O.L. The insert shows a perspective view of a Type C antenna.

is 45 degrees, in fig. 23. It can be seen that this confirms the results of the Forestport study. Smith did the design work for the 837 foot vertical radiator used on 173 kHz by the Voice of America at Munich, Germany.[16] In this case θ was made approximately 55 degrees, and twelve top loading cables 512 feet long were used. The measured base impedance was 29.5 + j135 ohms. The measured unattenuated field at one mile was 178.9 mv/m for 1 KW, which is quite good efficiency. Scaling this antenna for 2 mHz use, its height would be 73 feet, and the top loading guy cables would be 44 feet long. As may be seen from the measured base impedance, the vertical radiator is slightly over ¼ wave in electrical height, the reactance being positive. There have been other studies made on this subject, some of which are those by Monser[17] and Gangi.[18]

Resistances

Unless one goes to the amount of top loading used on the Munich antenna, the resistance will be low, and the reactance will be capacitive. As will be seen from fig. 18(A) the base resistance of an antenna shorter than about 0.15 wavelength is less than 10 ohms, and the shorter the antenna the lower the resistance, and the higher the negative reactance. Adding the optimum top loading to the Forestport tower, while it brought the reactance to zero (resonance) at 80 kHz, left the resistance at only 4 ohms. In other words, top loading is effective in reducing capacitive input reactance of a vertical radiator which is quite short, but it does not have much effect on input resistance, which remains quite low. This means that antenna system Q is quite high, and particular care must be paid to keeping losses to a minimum. Base input resistance includes losses, of course.

The reasons for this can be seen in fig. 24, which shows the composite resistance characteristics of a v.l.f. antenna which is extremely short electrically. The loss components R_g and R_d are considerable when compared to radiation resistance R_r. R_g is conductive resistance, which represents power lost in conductors and earth due to current flow. R_d is dielectric loss resistance, which results from hysteresis losses in insulators, nearby trees, vegetation, and wood or masonry. John Walter, one of our experts in the v.l.f. and l.f. field, has written an excellent paper on this subject.[19] It may thus be seen that it behooves the antenna designer to make the radiator as tall as possible, so that radiation resistance may be high

[16] Smith, *et al,* "Very High Power Long Wave Broadcasting Station," *Proc of IRE,* August 1954.

[17] Monser, G. F., Sabin, W. D., "Antenna Design for Maximum L. F. Radiation," *Electronics,* June 3, 1960.

[18] Gangi, *et al,* "Characteristics of Electrically Short, Umbrella Top-Loaded Antennas," *IEEE Transactions on Antennas & Propagation,* Vol. AP-13, No. 6, Nov. 1965.

[19] Walter, J. C., Capt. USNR (Ret.), "Practical Analysis of LF and VLF Antenna Resistance," *Naval Engineers Journal,* Feb. 1966.

enough to make the losses of little consequence when compared with it.

Folded Unipole

So far in this chapter we have been speaking of series fed vertical radiators. There is another way in which the efficiency can be improved, and that is by use of the folded unipole principle of feed.[20, 21] We shall not now go into the details of how a folded unipole works as that subject will be covered in a later chapter. Let it suffice to say that the folded unipole is the upper half of a vertical folded dipole, with the lower half replaced by a ground plane. If you are not using one on the amateur bands, you probably are using one as the driven element of your home TV antenna. They are very common in this application

The folded unipole feed principle may be easily applied to the short, top loaded vertical radiator. The transformer action of the folded unipole is used to give a more favorable input resistance than can be obtained with series feed. Another thing that the folded unipole feed does is to reverse the sign of the input reactance. The input reactance of a series fed tower shorter than ¼ wavelength is always capacitive. This means that a series loading coil (spoken of in high power as a *helix*) must be used to resonate the tower. With folded unipole input the feed point reactance is always positive. (Consider the tower and feed wire to be a shorted transmission line less than ¼ wave long. Its input reactance is positive.) Thus the folded unipole may be fed with a simple low-loss capacitive feed network, rather than through a loading coil or helix which is lossy. Also, because with the folded unipole feed the resistance will be stepped up by transformer action and the reactance will usually be lower, the Q of the antenna will be lower, and its bandwidth will be greater, than with series feed.

Studies of this type of feed have been made by the Naval Ordinance Laboratory, Corona, California in an effort to reduce the size of l.f. antennas for ship and shore use.[22, 23] A simple design which evolved is shown in fig. 25.

[20] Leonard, *et al*, "Folded Unipole Antennas," *IEEE Transactions or Antennas and Propagation*, Vol. AP-3, No. 3, July 1955.

[21] Monser, J. G., "Calculating FoldedUnipole Antenna Parameters," *Electronic Industries*, Jan. 1960.

[22] Walters, A. W., "Antenna Miniaturization," (Naval Ordnance Lab., Corona, Calif.) in "*A Decade of Basic & Applied Science in the Navy*," Supt. of Documents, U.S. Government Printing Office, Wash., D.C.

[23] Seeley, E. W., "Small Antenna Study," NOL, Corona, Calif. Unclassified Report. (Not available for public distribution.)

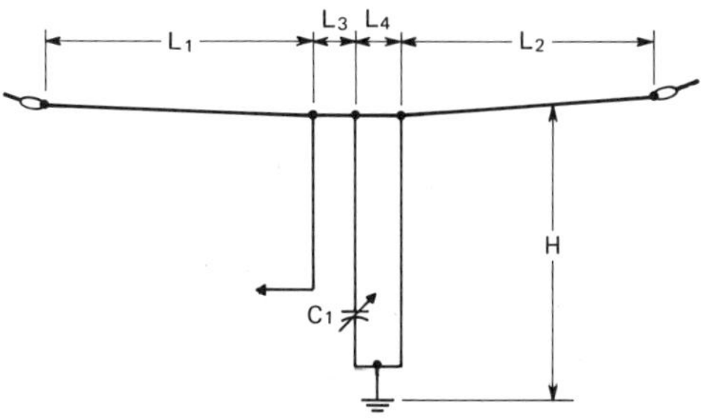

Fig. 26 — Configuration of the UG® antenna. The dimensions are given in the text.

Dimensions are shown in the accompanying table in terms of wavelength. It may be seen from the table that height H can be lower when 6 guys with skirt wire are used (Type *C*). Other dimensions are also smaller with Type *C*. For 2 mHz use, a Type *C* configuration would have the following dimensions:

H = 39.8 feet
S = 50.5 inches
L_1 = 39.1 feet
d_1 = 4.6 inches
L_2 = 55 feet
d_2 = 0.72 inches

For 4 mHz the above dimensions can be halved. The antenna can be fed with 50 ohm coaxial line.

Type UG

Another allied type of antenna which was developed several years ago is one called the Type UG.®[24] This type was devised to improve the feed point impedance and the bandwidth of some of the "inverted L" or "T" antennas used at l.f. The configuration of the Type UG is shown in fig. 26. Typical dimensions for 2 mHz would be:

L_1 = 25.0 feet
H = 25 feet
L_2 = 25.0 feet
L_3 = L_4 = 1 foot

The feed can also be applied to one end, if desired, as shown on fig. 27.

By study of these two figures it will be seen that the transformer action of the folded unipole has been made variable by the use of an extra downlead and capacitor C_1. In tuning up this antenna for a specific frequency, C_1 is varied until a favorable feed point impedance ($50+jx$) is seen at the feed point. This may then

[24] "Engineering Appendix-Type UG Antenna." Private communication from John H. Mullaney & Associates.

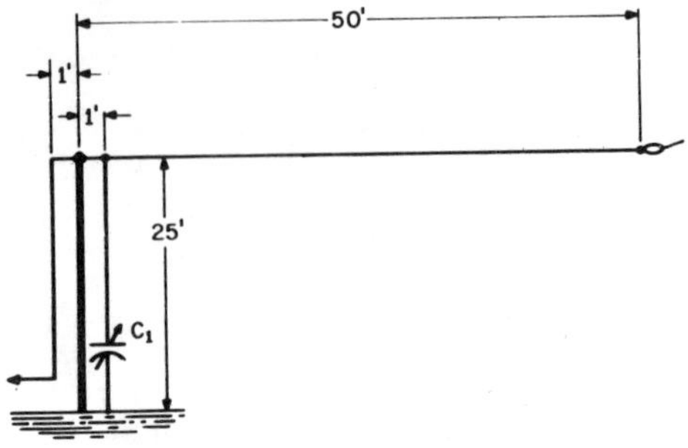

Fig. 27 — Configuration of an end fed UG® antenna.

be matched to the coaxial line by use of the matching techniques described in Chapter Three. One word of warning, capacitor C_1 should be a high voltage type if powers of 1 KW or so are used. Considerable current flows in this wire, and also in the grounded vertical wire, when the antenna is tuned up. In fact, the grounded vertical wire does not have to be a wire at all. It can be a metal mast such as a short tower or pipe, suitably guyed, and well-grounded at its base. Ground connections linking tuning unit, mast or tower, and C_1 should be heavy and well-made. I used an aluminum pipe mast for my end fed Type UG. The downleads should be rigidly mounted to the mast.

Another possible way of using the folded unipole feed principle for a six-band antenna will be described in a later chapter on the Mark IV DX Antenna. In this one, which is still in use, the 40 foot tower is used as a top-loaded vertical radiator on 2, 4 and 7 mHz, and the tri-band beam on top of it is the top-loading. On 2 and 4 mHz the tower is fed as a true folded unipole, with the feed line going to the top of the tower. On 7 mHz this tap point is too high, and therefor a feed rod is connected to the side of the tower at the 20 foot level. The schematics of these two feed arrangements for 2, 4 and 7 mHz are shown in figs. 28 (A)

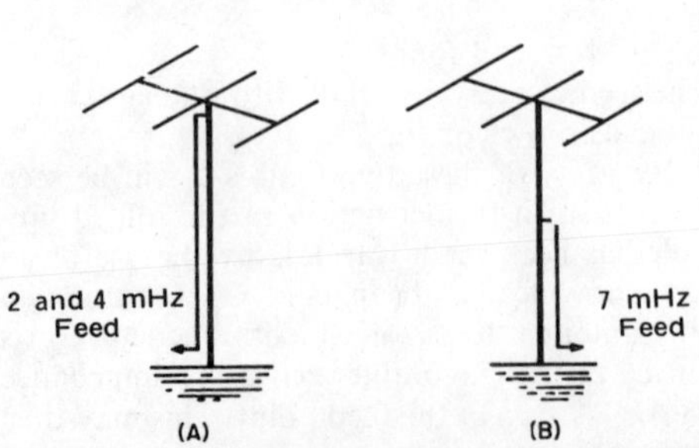

Fig. 28 — Feed systems used in the Mark IV DX Antenna.

and 28 (B) respectively. This antenna is ideal for the amateur who has a beam on a tower of reasonable height in his backyard, and who would like to get on 2, 4 or 7 mHz without putting up wire antennas.

NORD

There is yet another type of short vertical antenna which has recently come into use. This is the NORD®. The original NORD was designed for shipboard use, for radio broadcasting in the m.f. band in European waters. It has since been built for shore use at high power, in the l.f. band. To date, NORDs are in use at Naval Communication Stations in Japan, Guam, Philippines, Hawaii and Annapolis, Md., providing l.f. multi-channel transmitting capability. A schematic of the NORD is shown in fig. 29. A center tower is used, with its base grounded. The tower is fed by the folded unipole method, with a feedwire going to the top of the tower. The upper set of guys (three of them) is connected to the tower, and they extend out to three short perimeter towers, where they are secured to strain insulators. From these the cables drop down to tuning units which consist of capacitors.

For sake of clarity in fig. 29, the perimeter towers are not shown. The NORD is basically nothing but three over-coupled tuned circuits with a common element, the center tower. The purpose of the NORD for l.f. is to provide sufficient bandwith for multichannel teletype transmissions, while still retaining reasonable radiation efficiency. The electrically short series fed tower cannot do this at high power. Its Q is very high and its bandwidth is limited. Also, with top-loading, the top-loading guys are corona-limited due to the very high voltages which are developed on them. The NORD has no such problems. By adjustment of the capacitors C_a, C_b, and C_c, Optimum bandwidth and/or efficiency can be obtained, with reasonable input impedence being provided by the folded unipole feed. For Naval l.f. purposes we require bandwidth. For amateur c.w. and voice at h.f., efficiency is paramount. The NORD can fulfill both of these, by suitable adjustment, in a tradeoff of efficiency for bandwidth or vice versa. By going to the NORD at l.f. we have rid ourselves of the high voltage helix with its inherent losses, and its bandwidth limitations. The input coupling unit with the NORD can be nothing but low loss capacitors. In fact, in the high power NORDs vacuum capacitors are used throughout.

A full-size NORD, designed for 50 to 150 kHz., now in use, has dimensions as follows:

Height of center tower — 450′.

Height of perimeter towers — 150′.

Spacing of center to perimeter towers — 500′.[25]

For use on 2 and 4 mHz by amateurs, the NORD would not have to be so short electrically. It could have the following dimensions:

Height of center element — 30′.

Height of perimeter supports — 10′.

Spacing of center to perimeter — 33′.

This antenna would have excellent efficiency on both bands. Guy tuning capacitor values would have to be determined by experiment, but should be of at least 1500 mmf total capacity, adjustable in steps, with a portion variable. The input impedance at the feed point will be R + *jX* of a reasonable value, and this can be matched to coaxial line by a simple capacitor network. The antenna is tuned by connecting an impedance bridge to the input, then starting with the guy tuning capacitors at minimum they are brought up in value simultaneously until a good R + *jX* is obtained. It may be possible to obtain 50 + *jX*, in which case only a series capacitor will be required for matching to the line. Heavy ground busses must be used to link perimeter tuning units with the center tower.

There is another type of short vertical radiator that can be used at h.f., and this is the wound whip. These are made commercially by the Shakespeare Co. in South Carolina, and by others. A wire is wound around and imbedded in the outside of a fiberglass whip. The winding greatly increases the electrical length of the antenna. In fact, one might consider it to be a loading coil which has been stretched out to become *the* antenna itself. Whips of this type find great use on small boats, fishing craft, yachts, etc., for use with ship-shore radiotelephone in the 2 and 3 mHz band. Any amateur can easily build one by winding wire around a bamboo pole, wrapping the thing with waterproof tape and varnishing it. Such an antenna would suffice to handle several hundred watts, but for 1 KW or more I would not recommend it. Considerable voltage can be developed between turns in some places on the whip, and it could easily go up in smoke.

SLINKEY

For low power there is another similar type, fed in the folded unipole manner. This one can be made of a child's "Slinky Toy", the coiled steel spring which stretches itself and "walks" downstairs by itself. The spring can be stretched out on any suitable support such as a fiberglass tube, anchored in place, and fed like a folded unipole, against ground. Its length can be varied to produce almost any input impedance and degree of loading desired. Such an antenna, experimentally built, would be an ideal thing for indoor use, or sticking out from an apartment window, suitably supported. It offers endless opportunity for the enterprising amateur to surprise himself with the amazing results which can be obtained. Communications have been maintained between the Washington D.C. area and ZS stations and other DX locations using a short "Slinky" on 21 mHz s.s.b. The little thing never ceases to amaze me.[26]

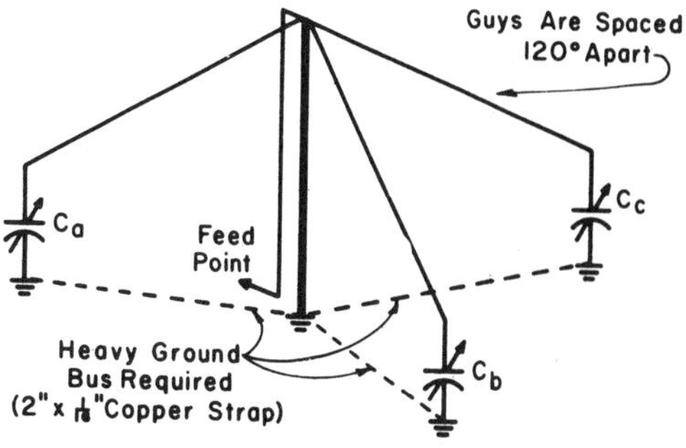

Fig. 29 — Layout of the NORD® Antenna. A heavy ground buss is needed, at least a 2″ × 1/16″ copper strap.

To sum up this chapter, there are many ways in which the serious amateur can use short vertical antennas and obtain good results. One need not be limited by space or height restrictions in getting on 2, 4 or 7 mHz. A major consideration is the ground system, however. Reference to figs 14 and 15 will show the importance of having sufficient radials of great enough length when the antenna is shorter than 0.15 wavelength. The shorter the antenna, the more important the ground system becomes in obtaining the highest possible radiation efficiency.

The types UG and NORD are covered by patents held by John H. Mullaney, and the "Slinky" technique is covered by a patent application pending. This does not prevent any amateur from building one for his own use, however. ■

[25] "NORD Antenna Model Report," Private communication from John H. Mullaney, Consulting Engineer, Potomac, Md.

[26] Private communication from John H. Mullaney, Consulting Engineer, Potomac, Md.

NOTE: The Slinkey antenna is now being marketed by Teletron Corp., Suite 300, Box 84, Kings Park, New York 11754.

DIRECTIONAL VERTICAL ANTENNA DESIGN

The author discusses directional antennas. Basic design principles are explained, a sample design is worked out, and methods of power division and phasing are shown. Amateur band configurations are covered.

IN past years there have been articles in amateur radio magazines on directional arrays using vertical elements. These have been of both types, driven arrays and parasitic arrays. For the most part these have been fixed arrays, with several elements which could be switched in or out to change the direction of the pattern. It is natural that a horizontal Yagi array, which can be put up in the air on a tower thus requiring practically no real estate, and which can be easily rotated to any direction, is in the majority. However, for the 2, 4 and 7 mHz bands, Yagi arrays are somewhat impractical because of their size. Thus there has been some interest in vertical arrays whose direction of radiation can be switched. There are several rather simple configurations of element spacing and phasing which lend themselves to this concept. It is the purpose of this chapter to explain how vertical arrays work, to show several simple configurations, and to sho㇒ how the elements of an array can be fed power of the proper phase and magnitude to produce the desired pattern.

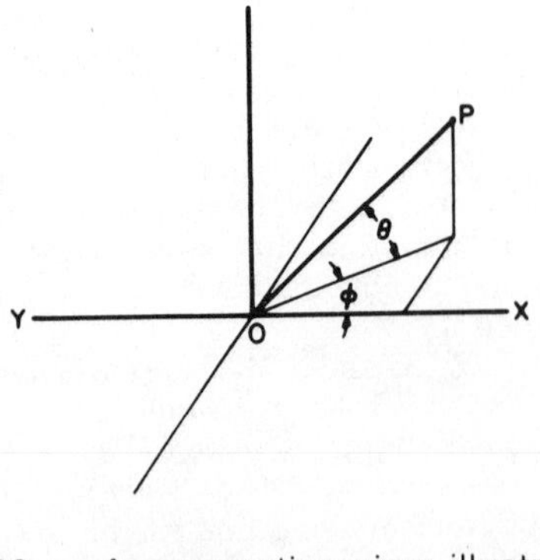

Fig. 30 — A perspective view illustrating the equation that permits the determination of the radiation to be expected in the direction of point P. In this drawing o represents the vertical angle of P, ϕ the azimuth angle of P, XY is the line of towers and the array is located at O.

M.F. Broadcasting Arrays

What is the purpose of a directional array? In the most widely used application, which is m.f. broadcasting, these arrays are usually used to produce nulls in the horizontal pattern (and sometimes also in the vertical pattern) to protect the service area of a co-channel station (and often an adjacent channel station) from interference. Coincidentally, the station's transmitter site is so located that the main lobe (or lobes) of the pattern falls over the station's desired service area, insofar as possible. Thus at m.f. the main purpose is protection of other stations from interference, not production of forward gain. Such an array often has several nulls in its pattern, one toward each station to be protected. The patterns are carefully designed, by application of proper phasing and current ratios to the vertical elements, to produce specific levels of field intensity in the nulls. The FCC always imposes specific levels of field intensity, which shall not be exceeded, towards the stations to be protected. Thus, it is the consulting engineer's job not only to design the pattern to fit these limits, but to assure himself that the array will be stable and remain in critical adjustment during changes in weather and ground conditions from season to season. Proper design of the phasing and branching networks enters into this latter consideration.

Amateur Use

For amateur use, however, we are not concerned with the absolute value of radiation in the nulls. Our main purpose is forward gain, usually in only one direction. Thus we can employ a more simple array of fewer elements, usually in-line, and content ourselves with obtaining forward gain and not worrying about depth of the nulls, nor about absolute pattern stability therein. Patterns can be more simple, with one main lobe (or possibly two). For reasons of propagation, such arrays for amateur use will usually be employed on 2, 4, or 7 mHz, although they can be used for the higher bands as well, instead of horizontal Yagis.

Review

It is well to run through some of the basic mathematics for directional arrays, so that we all know what we are talking about. Figure 30 shows a perspective view which illustrates the following equation, which gives the radiation to be expected in the direction of any point P in space from a simple two-element array. The line of towers is shown as XY. S is the spacing between towers. The horizontal or azimuth angle of point P from the line of towers is indicated as ϕ. The vertical angle of point P above ground plane is indicated as θ. The phasing of the currents in the two elements is shown as ψ, and R is the current ratio between the two elements. The field intensity at point P is thus equal to:

$$E = Kf(\theta)\sqrt{\left(\frac{1+R^2}{2R}\right) + \cos(S\cos\theta\cos\theta + \psi)}$$

This can also be written as:

$$E = Kf(\theta)\left(I_1 \angle 0^\circ + \frac{I_2}{I_1} \angle S\cos o\cos 0 + \psi\right)$$

K is a constant which is used to convert to mv/m, and which depends on the r.m.s. value of the pattern, the power used, and the height

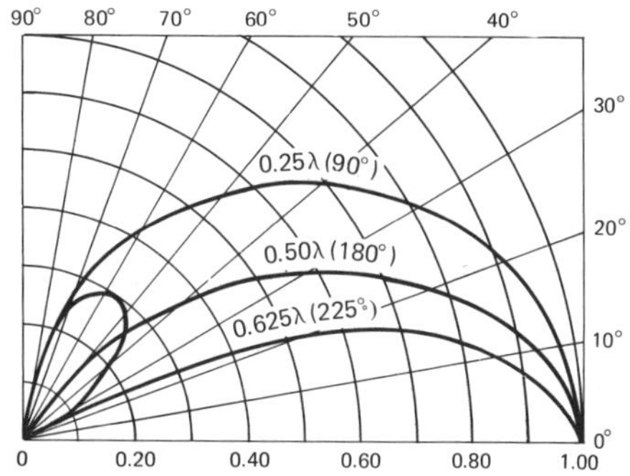

Fig. 31 — Plot of the factor f (θ) for towers of ¼, ½ and ⅝ wavelengths height.

of towers to be used. The factor $f(\theta)$ depends on the tower height G in degrees, and which is computed from the following formula:

$$f(\theta) = \frac{\cos(G\sin\theta) - \cos G}{(1-\cos G)\cos\theta}$$

For example, the $f(\theta)$ for a tower height of ⅝ wavelength (225° is as follows:

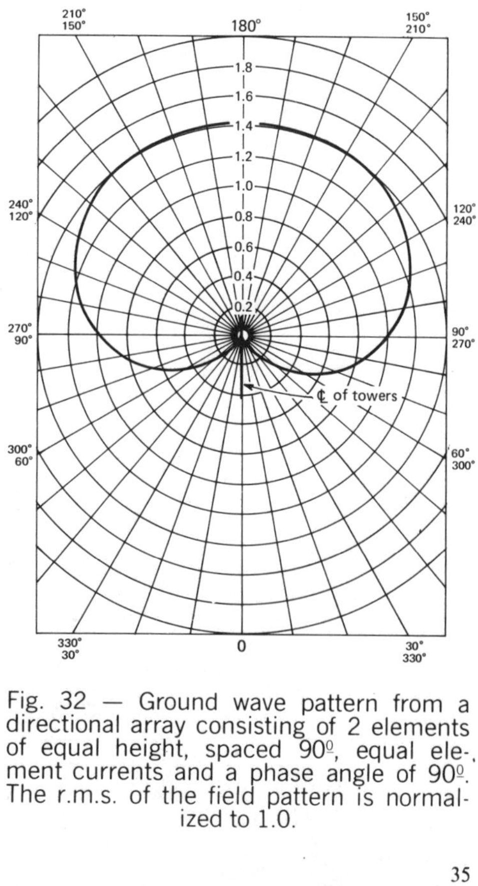

Fig. 32 — Ground wave pattern from a directional array consisting of 2 elements of equal height, spaced 90º, equal element currents and a phase angle of 90º. The r.m.s. of the field pattern is normalized to 1.0.

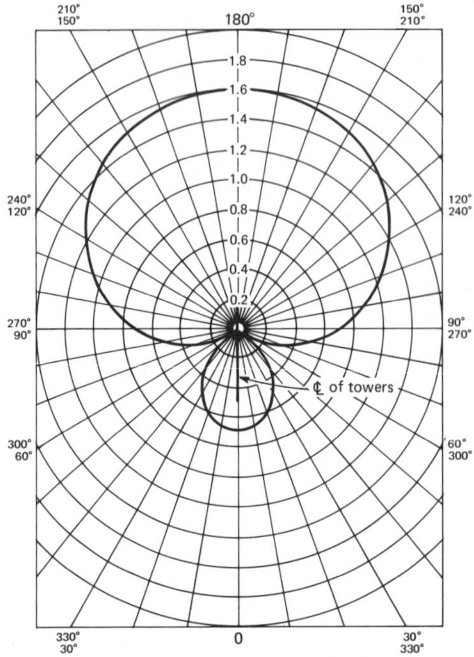

Fig. 33 — Ground wave pattern from a directional array consisting of 2 elements of equal height, spaced 90º, equal element currents and a phase angle of 135º. The r.m.s. of the field pattern is normalized to 1.0.

0°	$f(\theta)$
0	1.000
10	.883
20	.582
30	.219
40	—.119
50	—.260
60	—.292
70	—.250
80	—.138
90	.000

The minus sign is nothing to worry about. It merely shows that the radiation in the minor lobe is 180 degrees out of phase with that in the major lobe. Figure 31 shows this $f(\theta)$ plotted, along with similar plots for tower heights of ¼ wavelength (90°), and ½ wavelength (180°). It will be seen that this plot is the same as that of fig. 6. This is where the vertical patterns come from, right from the above equation for the value of $f(\theta)$.

Thus, for our two element array, we can compute the field intensity which will be radiated towards any point P in space. Now if point P lies in the horizontal plane, vertical angle θ is zero, and since cos 0° is unity, it falls out of the equation. Also, $f(\theta)$ is unity, and it falls out. Thus the horizontal or ground wave pattern equation is:

$$E = K\sqrt{\left(\frac{1+R^2}{2R}\right) + \cos(S\cos\theta + \psi)}$$

This can also be written as:

$$E = K\left(I \;\underline{/0^\circ} + \frac{I_2}{I_1}\;\underline{/S\cos\phi + \psi}\right)$$

If the currents in the two towers are made equal, the equations simplify to:

$$E = K\sqrt{1 + \cos(S\cos\phi + \psi)}$$

which can also be written as:

$$E = K\,(1\;\underline{/0^\circ} + I\;\underline{/S\cos\phi + \psi})$$

The azimuth angle of the null can be computed from the following relationship:

$$\cos(S\cos\phi_N + \psi) = -1, \text{ or}$$
$$S\cos\phi + \psi = \pm 180^\circ, \text{ solving for } \phi$$

The pattern is symmetrical about the line of towers, and there will be a pair (or in some cases two pairs) of nulls for a two tower array.

In actual practice, consulting engineers now use computers which are programmed for directional array design using these formulae. Given the limits of field intensity towards the stations which it is desired to protect, the com-

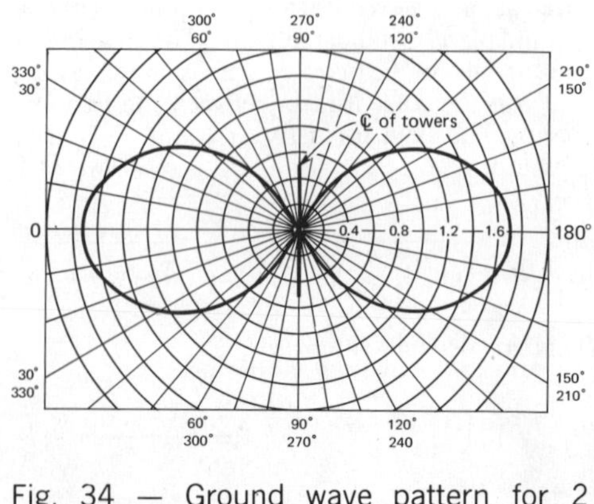

Fig. 34 — Ground wave pattern for 2 elements of equal height spaced 180º, equal element currents but with zero degrees phase angle.

puter can design an array whose pattern will fit those limits.

The pattern for a simple array of two towers whose spacing is ¼ wavelength (90°) and whose phasing is 90° is shown in fig. 32. Currents are equal. The figure is normalized to an r.m.s. of 1.0 for the pattern. Maximum radiation is only 1.42, which is a power gain of about 2, or 3 db.

Figure 33 shows a plot of radiation for a pair of towers spaced 90° but phased 135°. It will be noted that a small back lobe has appeared, and there are two nulls, symmetrically placed on either side of the line of towers. The voltage gain of this one is 1.6, and the power gain is 2.56, which is 4.08 db.

Figure 34 shows the pattern from a pair of towers spaced ½ wavelength (180°), and operated in phase. Note that the direction of the main lobes has shifted from that of the line of towers to a broadside direction. Voltage gain is 1.7, power gain is 2.89 or 4.62 db, and we have two nulls in the line of towers. If the spacing is increased beyond 180°, and the phasing maintained at zero, a small lobe will appear in each of the nulls in the line of towers, in this pattern. Thus there will be four nulls instead of two.

The maximum possible gain with two towers for a pattern whose main lobe is in line with the line of towers is a voltage gain of 1.725, or 4.74 db, and this occurs with spacing of approximately 30° and phasing of 165°. For the broadside pattern (two lobes), maximum gain is 1.82, or 5.2 db, and this results from a spacing of approximately 210° and phasing of zero degrees. (With this spacing there will be a very small minor lobe in each null.)

Multiple Towers

So much for two tower configurations. It will be seen from the above statements that gain is somewhat limited, and the lobes are quite broad, especially in the in-line cases. How can we increase gain? The answer is to increase the number of towers to three, or even four. It is not intended that this text go into all the various complicated arrays which can be formed. We shall simply state that in the case of an in-line array such as that whose pattern is shown in fig. 32, if a third tower is added at a spacing of 90° from one of the others, making it three in line with 90° spacing, the pattern of fig. 32 can be "squared," or multiplied by itself, to produce the pattern shown in fig. 35. The equation for such a pattern is:

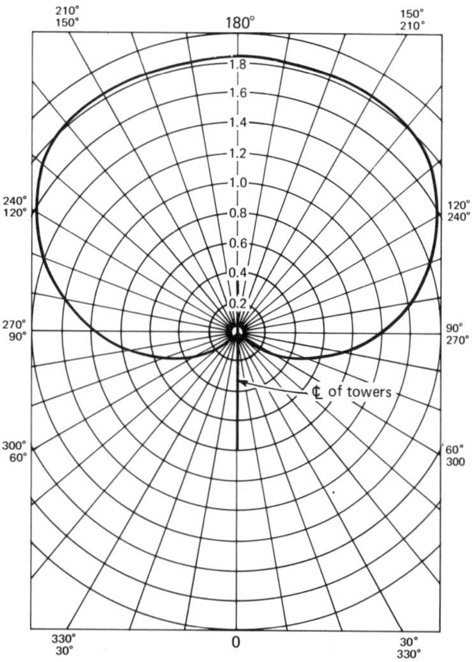

Fig. 35 — Ground wave pattern from a directional array comprising three elements of equal height, spaced 90° in line with $I_A = 1.0$, $I_B = 2.00$, $I_C = 1.0$, $\psi_A = 0$, $\psi_B = 90°$, and $\psi_C = 180°$. The r.m.s. of the pattern is normalized to 1.0.

$$E = \sqrt{[1 + \cos(S_2 \cos\phi_2 + \psi_2)] \times [1 + \cos(S_3 \cos\phi_3 + \psi_3)]}$$

For the in-line case, $\phi_2 = \phi_3$. The transition from the two tower case to the three tower case is shown diagramatically in fig. 36. Current in each tower is assumed to be unity. It will be seen that we start with two pairs of towers, each spaced and phased 90°, but we end up with three in line, whose phasings are entirely different, which produce the pattern of fig. 35.

Unsymmetrical Patterns

If the three towers are not to be in line, making an unsymmetrical, ϕ_2 will differ from ϕ_3. Or perhaps the spacing and phasing of the second pair will not be the same as those of the first pair, making an entirely different result. It can be seen, therefore, that there are an almost finite number of configurations, which produce all kinds of strange looking patterns. There is a very fine book[27], by Carl E. Smith

[27]Smith, Carl E., "Directional Antennas," *Cleveland Institute of Electronics.*

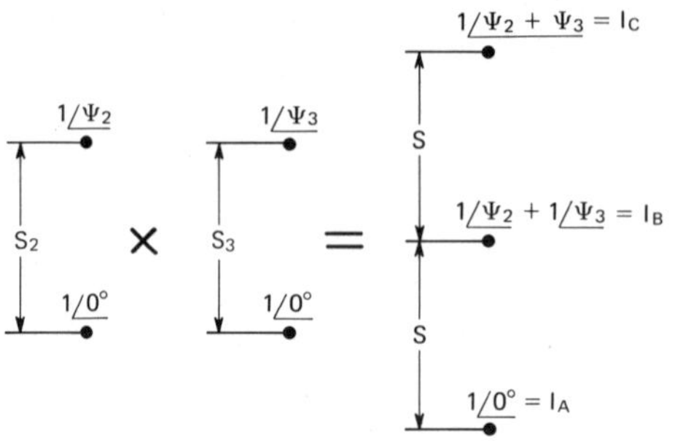

$\Psi_2 = 90°$
$\Psi_3 = 90°$
$S_2 = S_3 = 90°$

$I_A = 1.00\angle 0°$
$I_B = 2.00\angle 90°$
$I_C = 1.00\angle 180°$
$S = \angle 90°$

Fig. 36 — Multiplication of two equal tower arrays to form a three tower array.

which shows hundreds of possible configurations. The same author has written another book[28] which goes into the design of directional antennas in great detail, covering not only two and three tower cases, but also those with four or more towers. The mathematical principles for the more complicated arrays are the same as those for the simple arrays, but the computations are more laborious and time-consuming, unless one has a computer available.

To get back to our three element array of fig. 35, we see that the voltage gain is now 1.86, and the power gain is 3.46 or 5.4 db. By adding the third tower we have gone from a gain of 3 db in fig. 32 to a gain of 5.4 db in fig. 35.

For our maximum possible case previously mentioned, where we had a two-tower gain of about 4.74 db (the in-line case) we could add a third tower and realize a gain of about 9 db. In this case, and referring to fig. 36 for guidance, I_A would be $1.00/0°$, I_B would be $2.00/165°$, and I_C would be $1.00/—30°$. For the maximum possible gain in the broadside case, adding a third tower would result in about 10 db gain.

Field Intensities

In our discussion so far, we have normalized everything to 1.0 To obtain actual expected unattenuated field intensity at one mile from fig. 32 through 35, one must multiply by the factor *K*. For 1 kw into an array of ¼ wave elements, a value of *K* of 190 mv/m is reasonable. The FCC allows up to 7.5% power loss in the phasing and branching networks for m.f. broadcast directional arrays. When an array is tuned up and in adjustment, the proof of performance consists of making field intensity measurements on at least eight radials, and plotting these measurements on field intensity paper is described in Chapter Two. The unattenuated field at one mile is then obtained from these plots, for each radial. These values are then plotted on polar paper like that in figs. 32 through 35, and if the array is adjusted properly, the polar plot of measured values should conform to the designer's original plot.

Feeding

How are such arrays fed? It is a well known fact that there are mutual interactions between elements. What causes this? What can be done in the design of the phasing and branching networks to take these interactions into account and to compensate for them? First, let us examine the interaction itself. There is a very good explanation in a paper by G. H. Brown[29]. When two vertical (or any other kind) antennas are close together, there exists at antenna number two a voltage which is produced by the current flowing in number one. Thus a current flows in number two due to the current flowing in number one. This current is over and above the current which flows in number two due to its own connection to the source of power, and it bears a certain phase and magnitude relationship to the current in number one which causes it.

This phase and magnitude relationship is caused by the "mutual impedance" between the two antennas. The mutual impedance depends on the antenna heights and the spacing between them. Figure 37 is a plot of the resistance and reactance parts of the mutual impedance between two ¼ wave vertical radiators (referred to the current loop at the base of the antenna). Note that both resistance and reactance can be either positive or negative in sign. There are similar charts for other tower heights, both for towers of equal height and of unequal height, in Smith[28] and in Brown[29].

The array designer uses these charts in the initial design of the phasing and branching networks. Actually, in the field, the engineer can measure and compute the mutual impedance between two towers by measuring the change in self-impedance of one tower when the other is first open-circuited and then resonated to ground. He can then go to the second tower and repeat the process in reverse, if the towers are equal height, and he should get the same re-

[28]Smith, Carl E., "Theory and Design of Directional Antennas," *Cleveland Institute of Electronics*.

[29]Brown, G. H., "Directional Antennas," *Proceedings of the I.R.E.*, January 1937, pages 78-145.

sult (within a couple of percent). There is a formula[29] for this computation, but we shall not go into it here. The mutual impedance determined in the field by this method usually is quite close to that which can be read from a chart such as fig. 37.

Having found the mutual impedance either from the chart or by actual field antenna resistance and reactance measurements, the engineer can then determine the actual operating base impedance of each element of the array, when all the elements are connected to the system and handling power. This is something which cannot be measured by an ordinary radio frequency bridge, because the element or tower cannot be detached from the system for purposes of making impedance measurements. Detaching it destroys the effect of the mutual impedance and gives an entirely erroneous result. The operating impedance of a single element under power can only be determined by use of an s.w.r. bridge which will handle the power involved. One company, Delta Electronics of Alexandria, Va. makes one for m.f. use for powers up to 5 kw. This unit is called the OIB (Operating Impedance Bridge) and resistance and reactance dials are calibrated in ohms. This has become a very handy tool for the consulting engineer to use in the field. The operating impedance can be computed, however, by use of the following formulas:

$$Z_1 = Z_{11} + \left(\frac{I_2}{I_1}\right) Z_{21}$$

$$Z_2 = Z_{22} + \left(\frac{I_1}{I_2}\right) Z_{12}$$

Where Z_1 is the operating impedance of tower 1, Z_{11} is the self impedance of tower 1 with tower 2 floating, I_2/I_1 is the ratio of currents, and Z_{21} is the mutual impedance from tower 2 to tower 1. Z_2 is the operating impedance of tower 2, Z_{22} is the self-impedance of tower 2 with tower 1 floating, I_1/I_2 is the inverse of the current ratio, and Z_{12} is the mutual impedance from tower 1 to tower 2. Theoretically, Z_{11} and Z_{22} should be equal, and Z_{21} and Z_{12} should be equal, and in a very carefully designed and constructed array this will be the case, within a couple of percent, with equal height towers. However, during actual tuneup and adjustment, the engineer always measures and computes these factors, both for a check on his initial design, and to show up any unforeseen defects in the array.

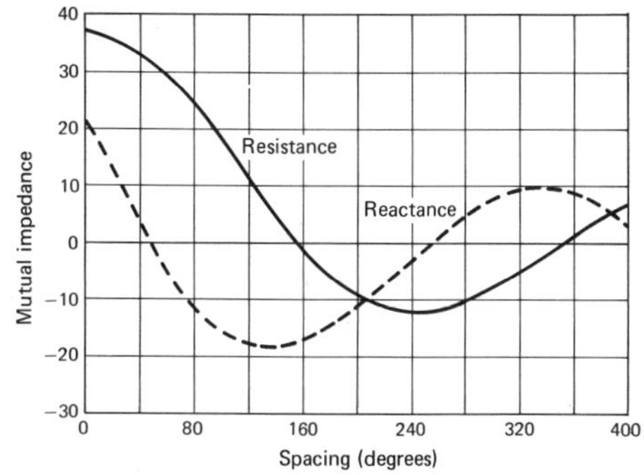

Fig. 37 — Plot of mutual impedance between two quarter wave antennas over perfect earth. It is referred to the current loop at the base of the antenna.

Coupling Networks

It can be seen from the above equations that for most cases the operating base impedances of the towers will be different by an amount dependent on the magnitude and phase angle of the mutual impedance, and on the phase relationship and magnitudes of the tower currents. Thus, the coupling networks must be designed to match those operating base impedances to the transmission lines being used. Designing the coupling networks to match the self-impedance of either tower would result in considerable impedance mismatch as well as an erroneous phase shift.

A typical two-tower phasing and branching network is shown in fig 38. The two transmission lines are tapped on a low-Q tank circuit, which is in turn tapped to provide a suitable load (usually 50 + j0 ohms) to the transmitter. This type of circuit is called "the jeep coil type," because of the fine adjustment coils marked "X." It is a very critical circuit to adjust, especially when three or more towers are being fed, because of the extreme interactions between individual adjustments.

A much better circuit in this respect is shown in fig. 39. This is called "the Ohm's Law type." It is much easier to adjust when three or more towers are used, because each tower is fed off a separate power divider coil, and the parallel combination of these, in series with L_S, is tuned by capacitor C_A to form a low-Q tank. The tank circuit's input impedance is usually set to be about 150 + j0 to 250 + j0 ohms, and this is matched to the transmitter, which usually likes to see 50 + j0 ohms, by a simple "T" network.

In the circuits shown in fig. 38 and 39, the

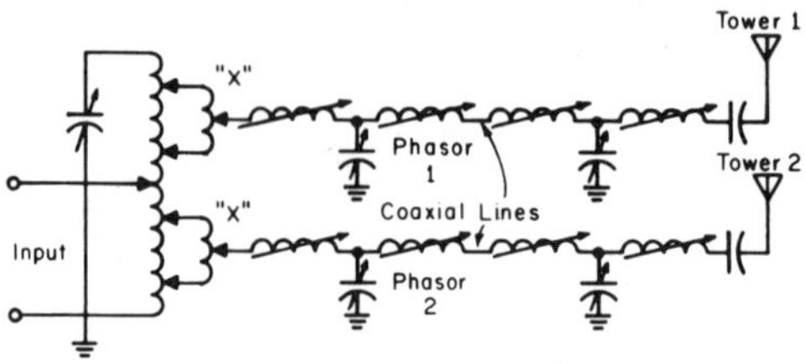

Fig. 38 — A typical two tower phasing and branching network called the "jeep coil type."

amount of phasing to be contributed by the phasing networks and the tower coupling units is computed; the required component values are then determined by calculation, and the whole thing is then wired up and tested under power (low power at first). The engineer makes the required adjustments to secure the desired phasings and current ratios, and checks the actual field intensity pattern by field measurements. Two-way radio is used, with a field car located in each null, calling in readings as adjustments are made, until the pattern is correct. S.w.r.'s on the lines are checked with the Delta Bridge, required impedance matching adjustments are made (which throws off the array adjustment somewhat), and the process starts all over again. Tuning up a two tower array can usually be accomplished in two or three nights of work (midnight to 6 A.M.), but a multi-element array with critical or deep nulls can take as long as three or four weeks of very tedious adjustments. By keeping a complete and careful log of all adjustments and their resulting field readings, the engineer can soon note trends, and know which way to go with the various adjustments, which always interact somewhat.

Amateur Use

While all of this is not applicable in detail to the amateur case, we have gone through it step by step to show the basic principles involved, so that you will have an understanding of all that goes into the design and operation of these arrays. Perhaps some of you will want to apply these principles to the design of your own arrays for 2, 4 or 7 mHz. If so, you will find Smith [27, 28] and Brown[29] invaluable. It might also pay to visit one of your local broadcast stations which has a directional array, and look over the circuitry and discuss its design with the chief engineer. Note particularly the type of power dividing or branching network used ("jeep coil" or "Ohms Law"), how the phasing is accomplished (positive or negative networks), how the tower coupling units are designed, how the tower lighting circuits are brought across the base insulators (choke coils or Austin transformers), how the ground system is designed, and its extent, and, in general, how carefully put together the system is. Most of them are quite well designed and built; they have to be, to pass FCC inspection.

Configurations

Now, for some simple configurations which lend themselves to amateur use, refer to figs. 32 and 34. For the array of fig. 32, the 90° phasing can be accomplished by making the coaxial lines 90° different in length. This is an old trick, often used, for phasing of any amount. A simple matching network is used at each tower, and the Ohms Law power divider is employed. See fig. 40 for a suitable circuit. For the array of fig. 34, if you want a broadside array which will fire in two opposite directions, the phasing is zero degrees. This is very easily accomplished by making the coaxial lines equal in length. In fact, the array is completely symmetrical, and the tower tuning units are alike. Again an Ohms Law network can be used.

For the maximum gain case mentioned previously, locate the towers 30° apart, make the coaxial lines about 135° different in length,

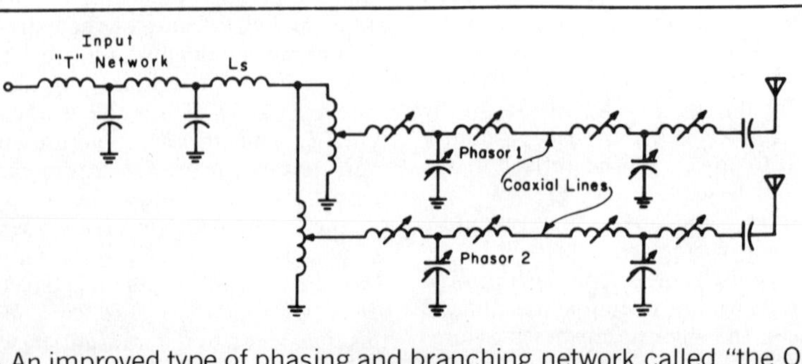

Fig. 39 — An improved type of phasing and branching network called "the Ohm's Law type."

and design the tower coupling units for the operating impedances involved. In this case, the effect of mutual impedance will be quite pronounced due to the close spacing, and more attention till have to be paid to s.w.r. on the lines, and to matching. In the two previous cases, perhaps some mismatch and s.w.r. can be tolerated, with only a kilowatt and RG-8/U cable. In the maximum gain case, however, more care should be taken, if the gain is to be realized. For the 210° maximum gain case, we again have a completely symmetrical array, and with the great spacing the effect of mutual impedance is not going to be so apparent. A bit of s.w.r. can be tolerated here.

Radials

Each tower or radiator should have its buried ground system. These should be joined by a copper bus running between tower tuning units. This can be made of copper strap, or three or four ground wires twisted together. Where ground radials overlap, they should be joined, as in fig. 41. The whole ground system should be one well-bonded unit, with no loose connections or unsoldered crossing wires. Refer to Chapter Two for a discussion of ground system design and in its effect on radiation efficiency. If I were going to build a directional array for 2, 4 or 7 mHz, I would lay down at least 32 radials around each radiator, with the ground bus extending into the building and connecting to the branching unit and the transmitter.

Tower Phasing

One matter I have not touched upon here is the measurement of the tower phasing. For the amateur this will not be too important, inasmuch as the array can be tuned for maximum forward gain (or maximum rear rejection) without much attention being paid to exact direction of the nulls. In the m.f. broadcast case, however, the phasing is required to

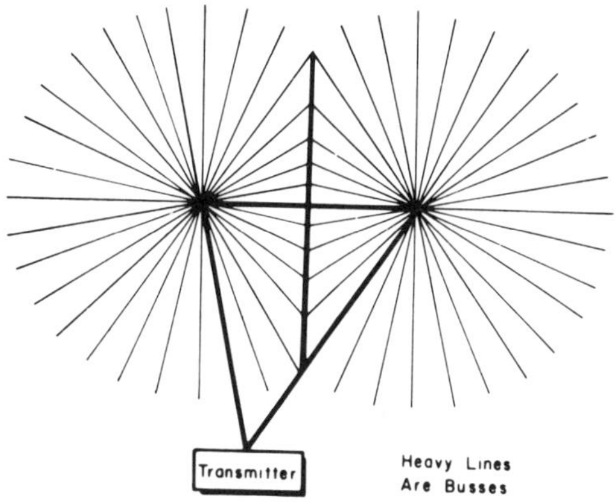

Fig. 41 — A 32 radial ground system for a two-tower array.

be logged, and maintained within two degrees. There is one common method of measuring it, and that is by means of a current sampling loop located on each tower. Each loop feeds a length of coaxial cable (usually RG-8/U), which is attached in the building to a phase monitor. This is a rather expensive piece of equipment which is designed to read phase difference between two currents. The sampling loop also provides a means of reading tower current remotely, and the remote meter can be calibrated to track fairly closely with the tower base current meter. Most phase monitors have provision for reading up to six towers. There has recently been introduced a simple current transformer which can be inserted in the lead from coupling unit to tower, to provide a remote indication. However, in the case of some self-supporting towers with wide bases, the tower input current will give a considerably different phase indication than will the pickup from the loop on the tower. After all, the actual tower current is what we are after. The pickup loop is usually mounted slightly above the base in a ¼ wave tower, or at the voltage

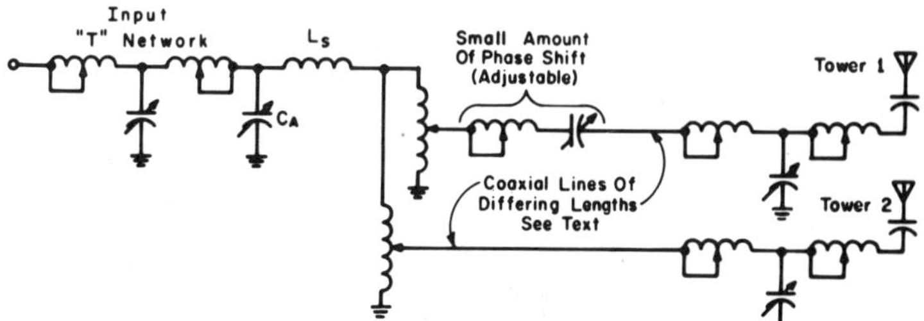

Fig. 40 — Circuit of an Ohm's Law power divider suitable for amateur use.

node (current loop) in the case of a tower taller than ¼ wave. The sampling loop is insulated from the tower, and its angular position is adjustable to vary the amount of pickup. The sampling line is brought down the tower on insulators, because it is usually buried from the base of the tower into the building. For the sake of easy reading of the phase monitor, it is customary to make all sampling lines of equal length, coiling up any excess in some out of the way place. This makes the monitor read correctly, without any "fudge factors." Perhaps some enterprising reader will build a phase monitor. It is not too difficult. Again, a visit to your local station may enable you to find a copy of a circuit digram.

2 mc Vertical Arrays

With the increases of power recently granted in some areas of the country on 2 mHz, and with crowded conditions on 4 mHz, use of vertical directional arrays might pay off. This leads me to think about possible use of such an array here on the west coast to protect the Loran coverage areas, while still delivering a strong signal to the interior of the country. This might be an interesting point to explore with the FCC — whether the FCC would permit one using an array to protect Loran areas to run increased power input over and above that normally permitted in his area. Such an array could be easily designed and tuned up, and with proper monitoring equipment it could be every bit as effective as one for the m.f. broadcast band. If any of you have any thoughts on this subject, I would be happy to hear of them.

If you have enough land available, you can put up a switchable directional array. For example, the array of fig. 34 can be used to provide coverage in six directions, by use of three elements spaced 180° on the corners of an equilateral triangle, and switched in or out by use of relays. The elements would be used a pair at a time. Other interesting configurations can be arranged, depending on the individual amateur's desired directions of communications. ■

THEORETICAL ASPECTS OF VERTICAL STACKING

An antenna composed of vertically stacked vertical elements produces low angle radiation without wasting power in high angle skywave lobes. In this chapter, the author covers the theoretical aspects of vertical stacking, and then describes some practical configurations.

IN the previous chapters of this book we have discussed vertical antennas composed of a single vertical element, whose height must be limited to ⅝ wave or less. For those who may be confused by the term "height," it is used to indicate the length of the vertical element or radiator. Figure 6 and Figure 31 show the vertical radiation patterns to be expected from antennas of several heights. The pattern for a ⅝ wave vertical radiator is shown. It has a large low angle lobe, and a small skywave lobe. These vertical patterns result from the computation of the factor $f(\theta)$ as shown in Chapter Five. If the computation is carried out for heights greater than ⅝ wavelength, it will be seen that the low angle lobe shrinks very fast, the skywave lobe increases rapidly, and the pattern becomes useless for long distance communications, as the height G is increased. Thus the statement that the height of the single simple element must be limited to ⅝ wave or less. Perhaps the statement should be qualified by saying that the vertical is most effective when producing low angle radiation for long distance communications. There are simpler antennas for short-hop, high angle skywave (such as a horizontal dipole close to ground).

Lowering The Angle

A look at the pattern of the ⅝ wave vertical shows, however, that some radiation is wasted in that high angle skywave lobe, and that the low angle lobe could be compressed even more and radiated at still lower angle, which is very desirable for long distance work.

Compression and lowering of the main lobe can only be accomplished by vertical stacking of in-phase elements. This is true whether the elements are horizontal or vertical. In the vertical element case, the array which results is called a vertical colinear array, since the elements are in line, or colinear. The vertical radiation pattern which results from a single two element colinear array in free space is shown in fig. 42. This array has a free space gain of 1.0 db over a dipole. The plot actually shows a vertical cross-section thru one side of the pattern. The whole pattern is actually a "dough-

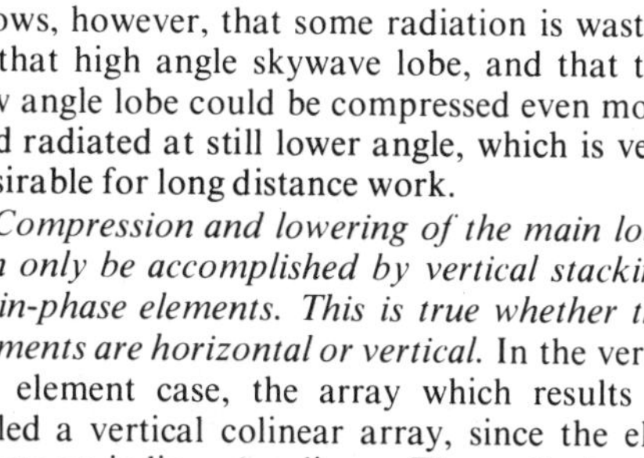

Fig. 42 — Configuration and polar pattern of a two element colinear in free space.

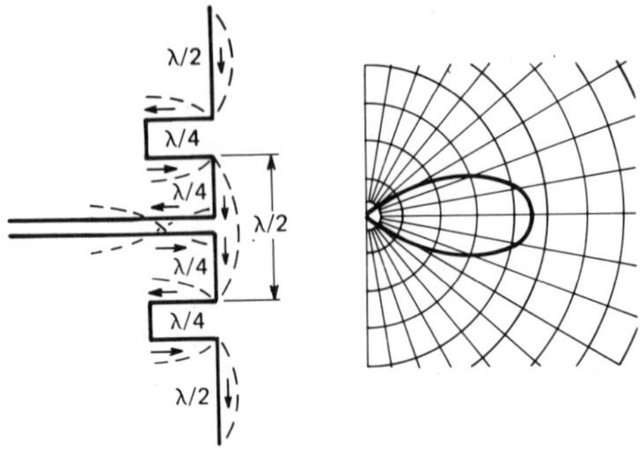

Fig. 43 — Configuration and polar pattern of a three element colinear in free space.

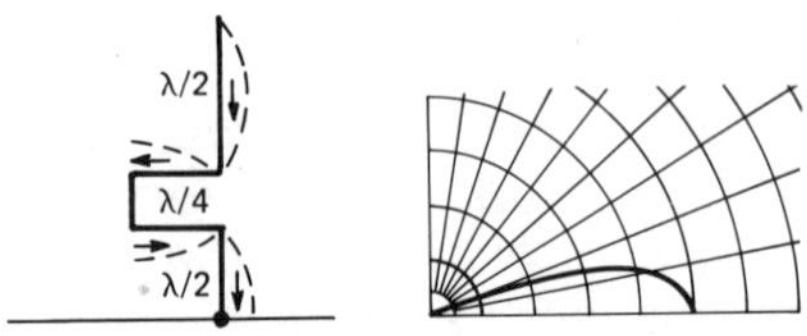

Fig. 44 — Configuration and polar pattern of a 1½ element colinear over ground.

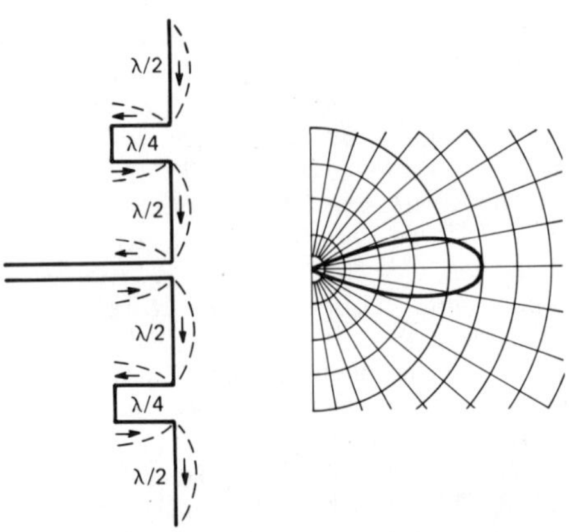

Fig. 45 — Configuration and polar pattern of a 4 element colinear in free space.

Fig. 46 — Configuration and polar pattern of a 2 element colinear over ground.

nut," in three dimensions. The two element colinear in free space, when cut in half, becomes a ½ wave element working against a ground plane. The ½ wave pattern in fig. 31 is the plot of this antenna's theoretical radiation over perfect ground. The effects of imperfect ground on the vertical pattern, and the resulting suck-in at very low angles were discussed in Chapter Two.

Figure 43 shows the pattern for a three-element colinear in free space. It has a gain of 3.2 db over a dipole. It also can be cut in half, and operated against a ground plane. In that case, however, one must be careful not to call it a ¾ wave antenna, for it is not that at all. It is a ½ wave element stacked above and operated *in-phase with* a ¼ wave element. Figure 44 shows this antenna and its vertical pattern over perfect ground.

If we carry the vertical stacking further, we can make a four element colinear array, whose free space pattern is shown in fig. 45. The gain of this one over a dipole is 4.3 db. As in the previous cases, we can replace the lower half by a ground plane, and then we have two ½ wave elements stacked, *in-phase*. This is not the same as a full wave antenna, whose two halves would not be in-phase. Figure 46 shows this antenna and its radiation over perfect ground.

Practical Colinear

From a practical standpoint the free space configuration of figures 43 and 45 is impossible to build and feed. It lends itself very nicely to textbook analysis, but it cannot be caused to produce the patterns shown, because of the presence of earth. The configurations which utilize a ground system as a part of the radiating system can be built, however, and the vertical patterns will conform closely to those shown in fig. 44 and 46, dependent on ground characteristics and frequency as discussed in Chapter Two. The effects of ground system design at h.f., and particularly the effects on low angle radiation, have been quite well documented in three reports by the U.S. Navy Electronics Laboratory.[30, 31, 32] The U.S. Navy uses

[30] "Ground System Effect on HF Antenna Propagation," Report 1346, *U.S. Navy Electronics Laboratory*, San Diego, 4 Jan. 1966 (Unclassified).

[31] "HF Extended Ground Systems: Results of a Numerical Analysis," Report 1359 *U.S. Navy Electronics Laboratory*, San Diego, 24 Feb. 1966 (Unclassified).

[32] "A Numerical Analysis of HF Extended Sector Ground Systems," Report 1430 *U.S. Navy Electronics Laboratory*, San Diego, 19 Jan. 1967 (Unclassified).

many vertical antennas as its shore radio stations, mostly in broadband configurations, and each one has its properly designed ground system. Briefly these reports show that greatly improved radiation at low angles will result when the number and length of ground system radials is increased, and that the improvement obtainable is a function of ground conductivity and frequency, with larger improvements being realized over poor ground (as would be suspected).

Franklin Antenna

It may be seen from the above configuration and pattern plots that, within practical limits, the greater the number of vertical inphase elements, the more the main lobe will be compressed and lowered, and the greater will be the gain. One such design, which carried the number of elements to a very high number, was the Franklin Antenna whose designer, C. S. Franklin, was very well known in the earlier years of radio communication in England. It was based on the principle that a vertical wire carrying uniform currents inphase over its length is the ideal radiator for producing the lowest possible angle of radiation. In the Franklin Antenna, as described in two of the references,[33,34] each successive ½ wave length of wire is folded back on itself in such a way that the radiation from its central part assists the radiation from its neighbors. This is shown in fig. 47 which depicts the current distribution on a Franklin Uniform Aerial. The exact dimensions in terms of wavelengths are in fig. 48. The major portions of the currents are inphase, and the resulting current distribution is almost the ideal of a uniform current. As a matter of historical interest, there is, in one of the references,[34] a very good photograph of a large Franklin Antenna array with reflectors at a Marconi station at Bodmin, Cornwall, England. This array consists of a series of vertical Franklin Antennas in line, with a reflector screen behind them. No gain figures are given for such arrays, but based on present designs of horizontal dipole curtain arrays of similar parameters,[35] the gain can be estimated at about 20 to 22 db. Such a Franklin array produces a narrow, low main lobe, broadside to the line of elements, by proper spacing and phasing of the elements.

[33]Williams, H. P., "Antenna Theory and Design, Vol. II," Pitman and Sons, Ltd.

[34]Ladner, A. W., Stoner, C. R., "Short Wave Wireless Communications," 5th Edition, John Wiley and Sons.

[35]Jasik, H., *"Antenna Engineering Handbook,"* McGraw Hill.

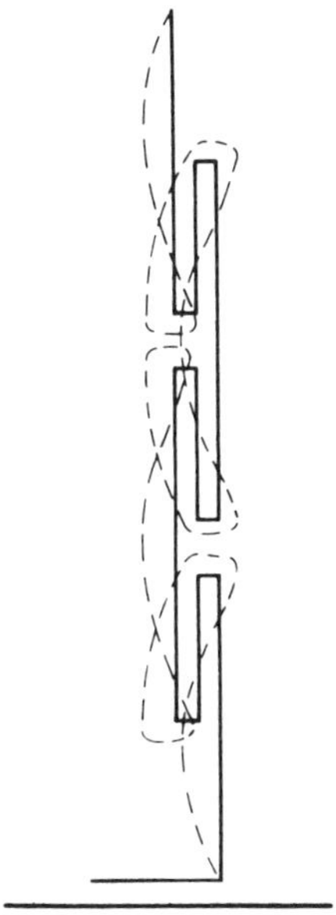

Fig. 47 — Current distribution on a folded Franklin Antenna.

Franklin Antenna Applications

You might well ask, if this array is so good, why do the Voice of America, the BBC, Radio Free Europe, *etc.* not use it in preference to the horizontal dipole curtain arrays? The answer

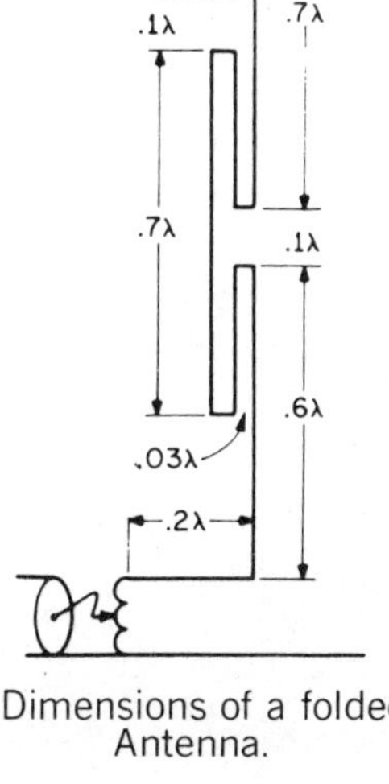

Fig. 48 — Dimensions of a folded Franklin Antenna.

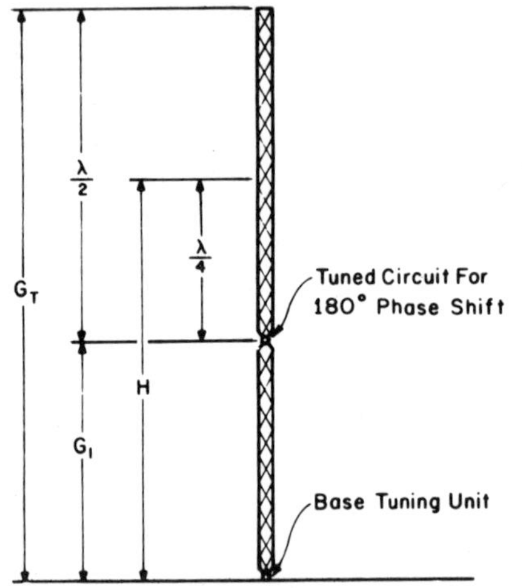

Fig. 49 — Sectionalized vertical radiator isolated by a tuned circuit. Length G is less than λ/2.

can be readily seen from the dimensions in fig. 48, which are very critical, and which make the Franklin Antenna *a single frequency* antenna. The horizontal dipole curtain array can, by use of broadbanding techniques in the design of the dipole elements and the feed system, be made to cover at least two adjacent high frequency broadcast bands, such as 7/9 mHz, 11/15 mHz or 17/22 mHz, with one antenna. Also, the Franklin Antenna is a bit more inconvenient to construct and rig, and it is difficult to isolate the vertical tower supporting structures, due to the vertical polarization. Thus the broadcasters and others engaged in

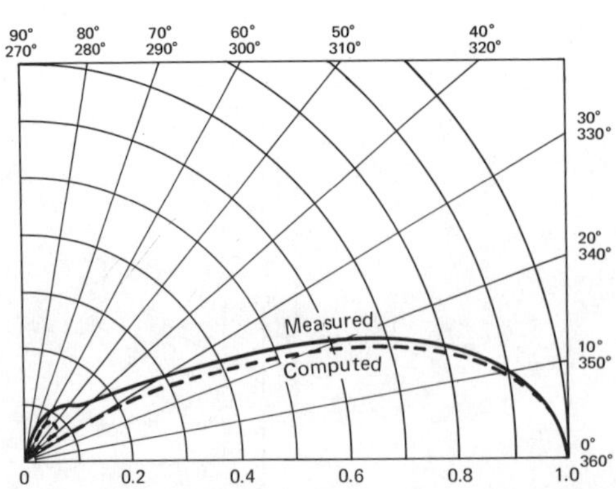

Fig. 50 — Vertical patterns for 300° sectionalized radiator at WHO, Des Moines. This is a plot by f θ where θ is the angle above the horizon. The measured pattern is normalized to 1.0 maximum value for θ = 0.

long distance communications have turned to broadband types such as curtains and rhombics, with horizontal polarization. It is not a matter of the polarization itself, but is a matter of operational and constructional convenience.

For v.h.f. or u.h.f. use, however, an amateur could easily bend up a Franklin Antenna out of stiff rod, and make it self-supporting and small in size, even though consisting of three or four ½ waves stacked. Figure 48 can be followed for design. If desired, the antenna can be fed against a ground plane, either by coaxial line directly (tolerating some v.s.w.r.), or via a suitable matching network composed of small components. Reference should be made to Chapter Three of this series for matching techniques. This will make a very fine v.h.f. or u.h.f. antenna, with low angle radiation and worthwhile gain.

Phasing Methods

The Franklin Antenna accomplished the necessary phasing between ½ wave elements by means of lengths of transmission line (although folded in a peculiar manner). One way of accomplishing the phasing is shown in figs. 44 and 46, by use of shorted ¼ wave stub lines. This is a bit impractical to rig mechanically, unless one has suitable supports for the vertical elements and for pulling off the wire stubs horizontally. There are two other ways which are more practical, and they are the use of tuned circuits and the use of coaxial sleeves. First, let's discuss the use of tuned circuits.

As you know, the ¼ wave shorted stub is the electrical equivalent of a parallel resonant circuit. Thus it can be replaced by a parallel resonant circuit, electrically. A vertical antenna consisting of ½ wave stacked over ¼ wave, or ½ wave stacked over ½ wave, or any other such combination, may be phased in this manner. The problem resolves itself into one of mechanical design, and weather-proofing the tuned circuit which must be mounted in the open on the mast or other support, or actually suspended in the antenna itself. One might think that such a tuned circuit would function as a trap, in the manner of traps used in multiband Yagi arrays, but it does not. The reason for this is that the length of the element beyond the tuned circuit is ½ wave resonant and it draws power through the tuned circuit. This is not the case with the length of the element beyond the trap in the Yagi. That short portion of the element in a Yagi cannot draw any power through the trap at the trap's resonant frequency. The trap in the Yagi acts as an isolating device, not as a phase reversing device.

There have been some really large-size examples of vertical antennas with tuned circuits for phase reversal. Several examples of this are shown in *Jasek*[35] in the chapter devoted to medium frequency broadcast antennas. The design of sectionalized vertical tower radiators is shown. In such cases, the tower is actually broken at the proper height by insulators, across which the tuned circuit is connected. Of course such a structure must be very carefully designed mechanically, and suitably guyed as it is quite tall. The reference shows the design of a sectionalized antenna of overall height of 300° used at WHO, Des Moines. The upper 180° (½ wave) of the tower is actually isolated from the lower 120° by insulators, and fed through a phase reversal tuned circuit mounted up in the tower. The vertical radiation pattern of such an antenna is given by the following equation:

$$f(\theta) = \frac{2\cos(90\sin\theta)\cos(H\sin\theta) + \cos(G_1\sin\theta) - \cos G_1}{\cos\theta}$$

where H and G are defined in fig. 49.

It may be seen that this is similar to the equation for $f(\theta)$ given in Chapter Five for a simple single vertical element. The vertical pattern of the WHO sectionalized tower is shown in fig. 50. There have been other such high power installations for m.f. and l.f. broadcasting in this country and in Europe. The tuned circuit must be built of suitable components and mounted in a weatherproof enclosure, as considerable r.f. voltage will exist across it.

Coaxial Sleeve

The other way of obtaining the required 180° phase shift is by the use of a coaxial ¼ wave sleeve. This is shown in fig. 51. The Mark III DX Antenna described in a later chapter is an example of a design of this type for use at h.f. The design of antennas of this type is covered in detail in several of the references[33 36]. Because of the mechanical problems connected with the design and mounting of the sleeve, such antennas are usually used for v.h.f. and u.h.f. where they can be physically small in size and easily supported. Some are even self-supporting, easily mounted on a mast. A coaxial sleeve on the Mark III was simulated by means of a wire cage, and it worked fine. Several readers did not quite have faith and confidence in the cage, and asked if they could build the sleeve out of sheet metal. I

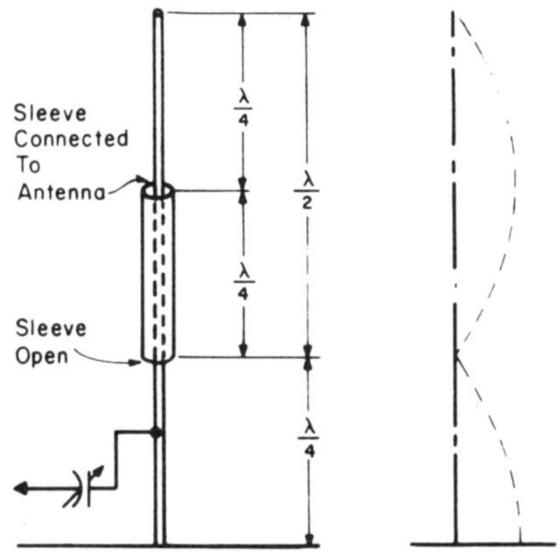

Fig. 51—A half wave over quarter wave in phase, with sleeve for phase reversal and its current distribution.

was a bit dubious as to their ability to install and support such a thing, and to maintain it in the face of problems of wind resistance, weight, *etc.*

Rigid Coax

Figure 52 shows a suitable design for v.h.f. or u.h.f. using rigid coaxial line. Several sleeves are used over the outer conductor of

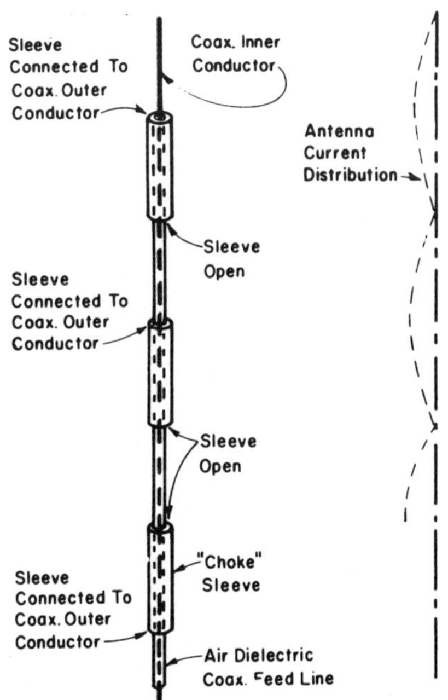

Fig. 52—Suitable design for a v.h.f. or u.h.f. colinear using rigid coax and sleeves as described in the text.

[36]King, R. P. *et al*, "*Transmission Lines, Antennas and Waveguides*," McGraw Hill.

the line, for phase reversal. The lower sleeve, it will be noted, is mounted with its upper end open, and is shorted to the coaxial line at the bottom. It acts in this manner as a detuning sleeve or "choke," preventing antenna currents from flowing any further down the outer conductor of the coaxial line. No ground plane is required in this case.

The ¼ wave coaxial sleeve is really nothing but a ¼ wave shorted section of transmission line, configured coaxially *around* the vertical radiator, instead of being made of wire and pulled off horizontally from it. It is quite a simple and effective device. There are, however, certain design precautions which must be observed. For maximum radiation it is necessary to individually tune (adjust the length of) each antenna section and each sleeve. In the case of the ¼ wave wire stub, this is quite easy. In the case of the coaxial sleeve, the outer surface (nominal ¼ wave) cannot be made shorter than the inner surface. A shorted stub of air dielectric coaxial line of ¼ wave electrical length will be very close to an actual physical ¼ wavelength. However, considering the outer surface of the sleeve only as an antenna, because of its considerable diameter and low L/D ratio it should be shorter than a physical ¼ wave to be ¼ wave resonant. Thus an apparently irreconcilable fact exists, which is that a ¼ wave section of sleeve cannot be ¼ wave long, yet it should be for proper phase reversal. To assure satisfactory operation under this condition, the inner diameter of the sleeve should be much greater than the outer diameter of the antenna conductor itself.[36]

A vertical colinear array should be so configured that it presents a suitable feedpoint impedance for connection to coaxial line either directly or through a simple matching network. One would not want to attempt to feed the bottom end of a ½ wave section directly, for example, for the impedance would be very high and hard to match. Bottom section of ¼ wave or ⅜ wave, however, are of reasonable feedpoint impedance and can be easily matched. In the Mark III a ¼ wave bottom section was used which was grounded and fed by the "gamma match" method as shown in fig. 51. Thus the antenna should be designed to consist of a number of ½ wave sections, over a bottom section of something less than a ½ wave for ease in feeding. If the antenna is to be self-supporting as was the Mark III, the grounded base design is an excellent idea, for it eliminates the problem of base insulation. With the coaxial sleeve design, other circuits such as lighting lines can be run right up inside the mast itself without worrying about isolation.

As with the Franklin Antenna, the tuned circuit isolation method and the coaxial sleeve method are both frequency sensitive. Since they depend on tuned circuit resonance and shorted stub resonance for phase reversal, a frequency change of more than five percent or so would be expected to degrade performance and cause considerable v.s.w.r. at the feed point.

Five percent would be adequate for coverage of the 7, 14 and 21 mHz bands, but it would not be adequate for the entire 28 mHz band, nor could one cover from 3.5 to 4 mHz without readjustment of tuned phase reversal circuits or stub lengths. To sum it up, the vertical colinear array, since it depends on accurate ½ wave phase reversals, is not a multi-frequency antenna. It is excellent for a narrow band of frequencies, but it is not easy to readjust for frequency changes. ■

BROADBAND VERTICAL ANTENNA DESIGN

The author describes several of the broadband types of vertical antennas which have been used by commercial and military services. Design information is given, and practical configurations are shown.

UP to now the vertical antennas discussed have been designed for specific frequencies or narrow bands of frequencies and the antennas had to be retuned for a QSY of any magnitude. Antennas of this type are usually used by amateurs, whose frequency allocations consist of relatively narrow bands in the spectrum. There has not been much material in amateur literature in the past regarding broadband types of vertical antennas. I shall now describe some broadband verticals and show how they can be applied to amateur requirements and additional types will be discussed later.

First let us examine some of the basic properties which tend to make a radiator broadband. It was described earlier how a structure's bandwidth is affected by its L/D ratio, with a shorter, fatter structure having a wider bandwidth (within certain s.w.r. limits) than a tall, thin one. The concept of a radiator being considered as an open-ended, lossy transmission line, extending into space, was also mentioned. It is well known practice to construct tapered transmission lines[37] to effect a smooth transition of impedance without creating standing waves. A tapered (or conically shaped) antenna may be considered as an attempt to effect a smooth transition between the impedance of a source and the impedance of space.

The Conical Antenna

A conical antenna may be considered as a special case of a horn antenna[38] commonly used at microwave frequencies. Such devices may be scaled up in size and used at h.f. The

[37]Laport, E. A. *"Radio Antenna Engineering,"* McGraw-Hill.

[38]Radio Research Lab, Staff, Harvard University, *"Very High Frequency Techniques,"* McGraw-Hill.

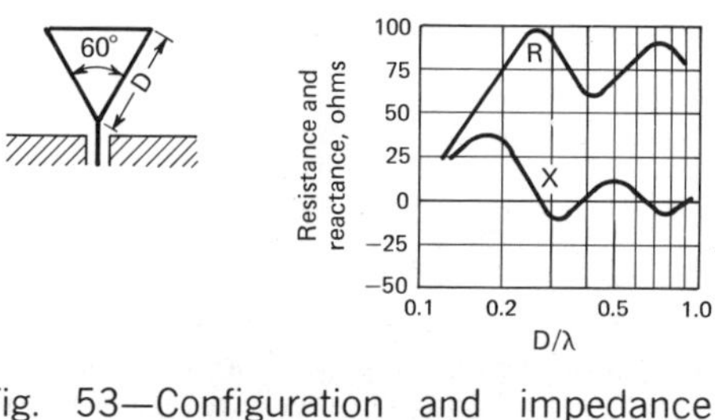

Fig. 53—Configuration and impedance plot of a conical antenna.

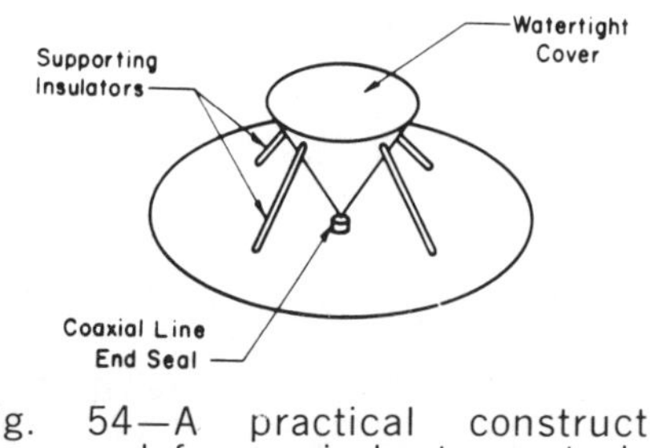

Fig. 54—A practical construction approach for a conical antenna at u.h.f.

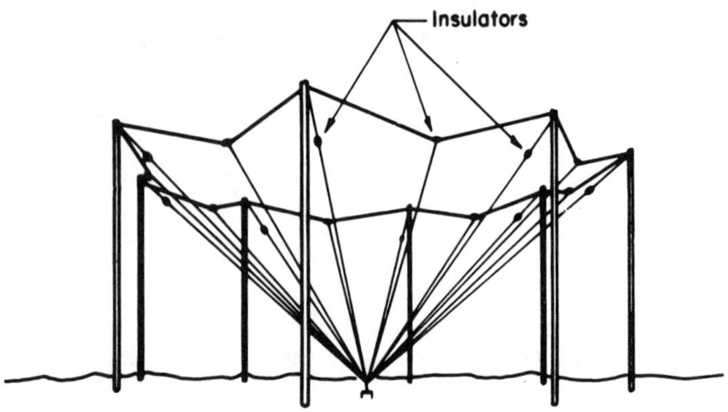

Fig. 55—A h.f. inverted cone consisting of 16 wires. The ground radials are not shown.

reference[38] shows results of model studies made of several conical configurations. Figure 53 shows the input impedance plot for a typical conical radiator. It will be noted that the cone is fed at its apex against a flat ground plane in this configuration. This type of antenna is quite easy to build for u.h.f. use, where the cone can be made of sheet metal and supported on three or more insulators, as in fig. 54. The lower frequency limit of this antenna is determined by the length of the side of the cone. As may be seen from fig. 53 it should not be less than ¼ wavelength, for good s.w.r. The high frequency limit is determined by the practical mechanical considerations of construction of the feed point. Due to its size at h.f., this type of antenna is rather difficult to construct. Some have been built, however, using closely spaced wires to form the cone, supported by a number of wood poles. Such a configuration is shown in fig. 55. A radial ground system is used as the disc.

The Discone

It is quite simple to invert the whole structure shown in fig. 55. This type of configuration is called a "discone." It was devised by Kandoian[39] [40] over 25 years ago. Actually the practice of using a cage of wires to form a broadband antenna is much older. The discone consists of a disc mounted over a cone whose apex connects to the outer conductor of a coaxial feedline. The disc is fed by the inner conductor of the coaxial line. Figure 56 shows the configuration of the discone. As in the case of the cone described in the preceding paragraph, the side of the inverted cone should be ¼ wavelength at the lowest frequency of operation. It has a useful frequency range of at least 10 to 1. For u.h.f. the cone can be made of sheet copper or brass, and the outer conductor of the coaxial line can be soldered directly to it

[39]Kandoian, A. G. *"Broadband Antenna,"* U.S. Patent 2,368,663, 6 Feb. 1945.

[40]Kandoian, A. G., "Three New Antenna Types and Their Applications," *Proceedings of I.R.E.*, Feb. 1946, p. 70W-77W.

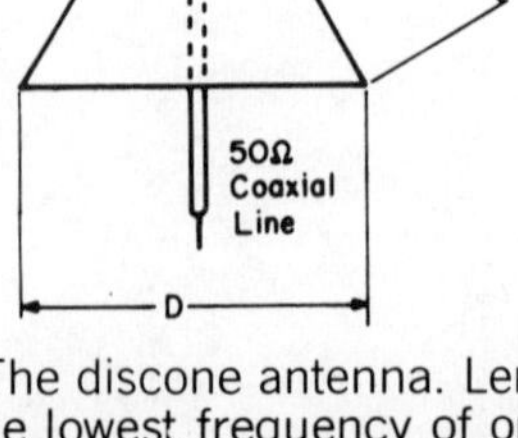

Fig. 56—The discone antenna. Length **D** is $\lambda/4$ at the lowest frequency of operation. The inner conductor of the feedline connects to the disc and the outer conductor connects to the cone.

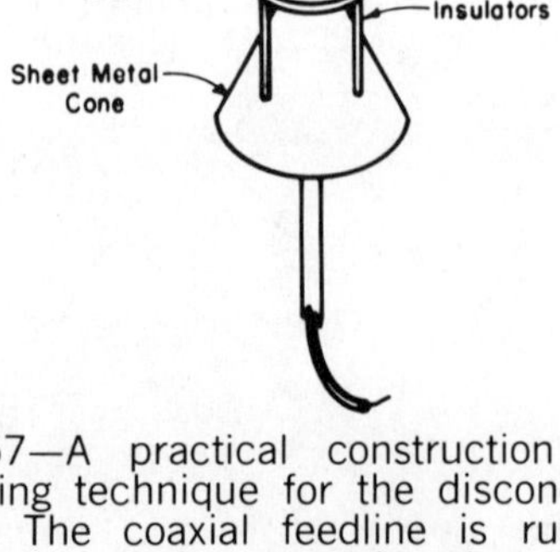

Fig. 57—A practical construction and mounting technique for the discone antenna. The coaxial feedline is run up through the supporting pipe.

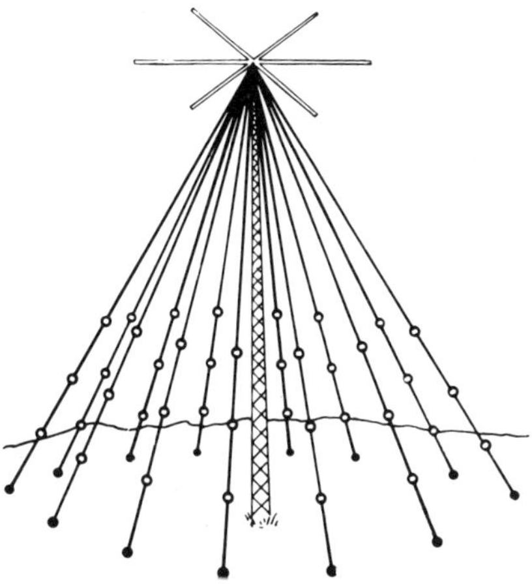

Fig. 58—An elevated discone antenna for h.f. use.

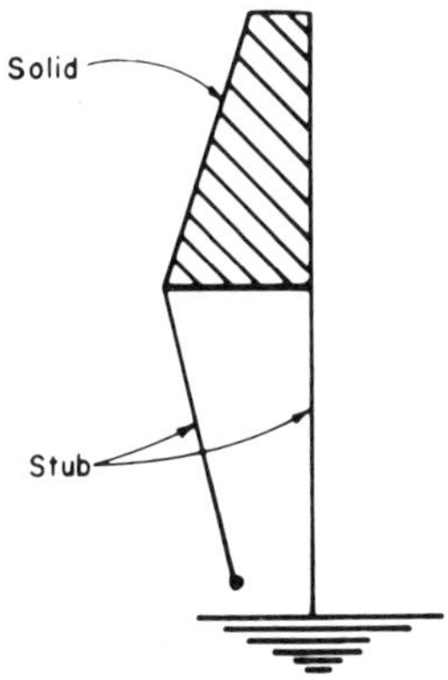

Fig. 60—View of a slice through half of a conical monopole to show the concept of a solid top with the bottom half a short section of transmission line or stub. In actuality the structure is simulated by wires and metal spreaders.

at its apex. The disc can be made of the same material, and supported on three or four fiberglass insulators as shown in fig. 57. The whole antenna can easily be mounted on a pole or other suitable support. Such an antenna would be excellent for use on the 50, 144, and 220 mHz amateur bands. In such a case, the side of the cone should be about ¼ wavelength at 50 mHz. The diameter of the disc should be 0.7 times the length of the side of the cone.

At h.f., the discone can be built using closely spaced wires to simulate the surface of the cone. The disc can be simulated by a structure consisting of six or eight spreaders, with wires connected between them. As in the case of the u.h.f. type, the h.f. discone can be mounted on a mast or tower. In fact, the configuration lends itself very nicely to tower use, for the wire elements comprising the cone can be used as guys for the supporting tower. This type of elevated discone is shown in fig. 58. The guys are broken with insulators, the upper portion of them comprising the cone.

As might be expected, the vertical pattern from this antenna, supported above ground, will be modified somewhat by ground reflections, exhibiting some lobing as shown in fig. 59. The low angle radiation, however, is quite good. When elevated as shown, this antenna requires no buried ground plane.

Conical Monopole

There is another type of antenna which has come into wide use by the military services. This is the conical monopole antenna, which is a broadband type useful over a frequency range of about 4 to 1. There have been several articles about this antenna in past amateur magazines,[41] [42] This antenna, however, was not developed by either of those authors, nor by the commercial companies with which one of them has been associated. The conical monopole was developed by Mr. M. I. Leppert of the Naval Research Laboratory, to give credit where credit is due. It was developed to fill an operational requirement for a broadband coaxial-fed antenna of superior characteristics for both shipboard and shore use.

The conical monopole has a simple mechanical design, short vertical height, good broad-

[41] Stroup, L. A., "The Conical Monople," *CQ*, Jan. 1966, p. 59.

[42] Pappenfus, E. W., "The Conical Monople," *QST*, Nov. 1966, p. 21.

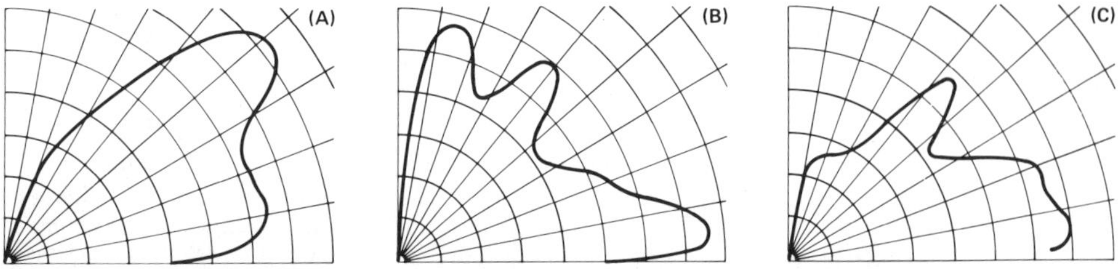

Fig. 59—Typical elevated discone patterns. The exact patterns will depend on the height above ground and the frequency range as well as ground conditions. (A) At lower frequency limit; (B) At midband frequency; (C) At high frequency limit.

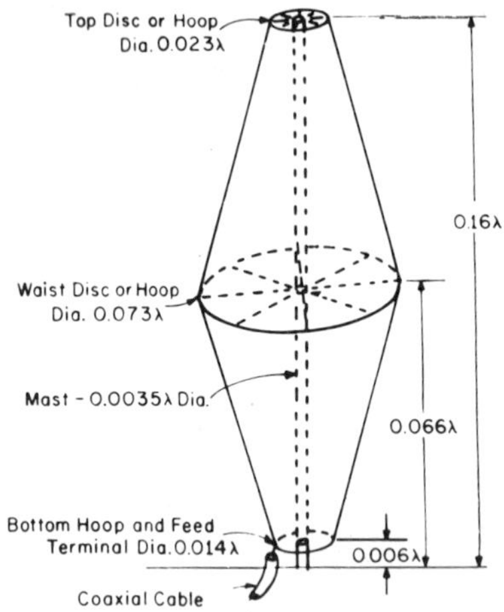

Fig. 61—Conical monopole configuration and dimensions. The dimensions are in wavelengths at the lowest frequency of operation.

band impedance characteristics, good radiation patterns and efficiency, and is capable of operating with a grounded mast or tower as its center support. The antenna consists of two truncated cones mounted base to base. The upper cone may be considered to be a solid, whereas the lower one is hollow. The inner surface of the cone and the supporting mast or tower form a short section of transmission line which is, in effect, connected across the terminals of the antenna formed by the outer surfaces of the two cones. Electrically it is a stubbed conical antenna, with the stub being internally contained within the structure. This basic concept is shown in fig. 60.

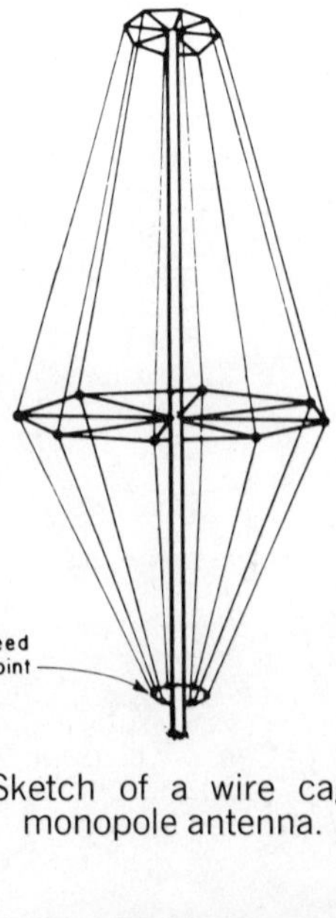

Fig. 62—Sketch of a wire cage conical monopole antenna.

The dimensions of the antenna developed by Leppert are shown in fig. 61. For practical purposes, the discs or hoops are simulated by means of metal spreaders, and the outer cone surfaces are simulated by wires, as shown in fig. 62. The references show some mechanical details of construction of antennas of this type.[41] [42] As with the discone type, this antenna can be built around a center tower or mast, using portions of the guys as the active cables of the antenna. The bottom loop or ring is pulled down by means of strain insulators.

The vertical patterns for this antenna are shown in fig. 63, for frequencies at which the antenna is 0.16, 0.50, 0.58, 0.64, 0.80 and 0.96 wavelengths tall. As may be seen, the pattern exhibits somewhat the same characteristics as that from a uniform cross-section vertical radiator. It consists of one major lowangle lobe for heights up to about 0.58 wavelength; it then starts to split and the lower lobe shrinks while the upper one grows until, at 0.80 wavelengths, there is only one high lobe. Then, at 0.96 wavelengths, the lower angle lobe appears again. It is for this reason that I stated previously that the antenna is useful over a 4 to 1 frequency range (heights from 0.16 to 0.64 wavelengths). With regard to s.w.r., it is useful over a frequency range of 6 to 1. Its s.w.r. is better than 2.7 to 1 over this range, relative to 70 ohms.

As with all grounded base antennas, this one must have a radial ground system. At naval shore stations we use ground systems of at least 120 radials a minimum of ½ wavelength long at the lowest frequency of the antenna. Of course, on ships, the metal hull and the sea water serve as the ground plane, a much better one than can ever be found at a shore station. The antenna is fed directly from 70 ohm line. When using 50 ohm line it is necessary to use a transformer section which consists of 0.196 wavelengths (at the antenna's lowest frequency) of 70 ohm line, inserted between the antenna and the 50 ohm feed. The s.w.r. will be within 2.7 to 1 over the 6 to 1 frequency range in this case also.

The Discage Antenna

As a further development for military use, the discage antenna combines the discone and the conical monopole in one structure to cover the entire range of 2 to 30 mHz. The idea orig-

[43] "Composite Discage Antenna Developed for 2 to 30 mc/s Band," *Naval Electronics Laboratory, Report No. 1504,* 8 August 1967 (Unclassified).

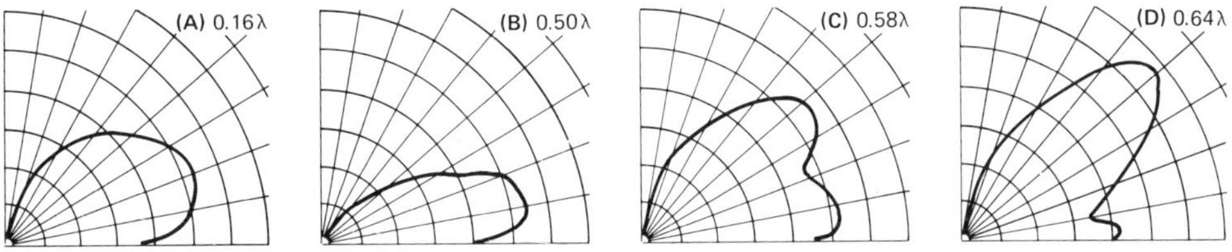

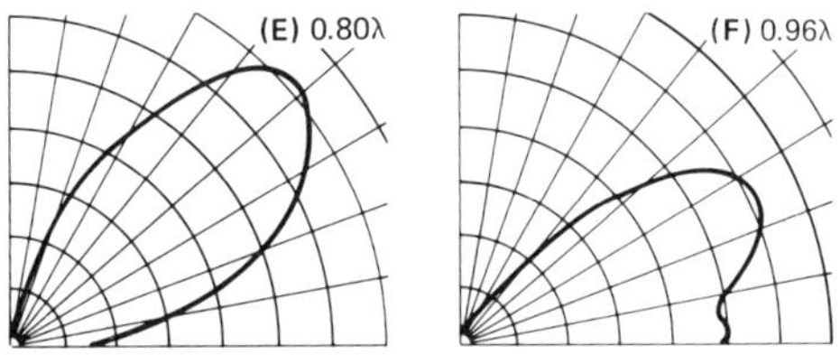

Fig. 63—Vertical patterns for the conical monopole at several heights.

inated at the Naval Electronics Laboratory in 1954.[43] Many versions of this type of antenna were developed for both ship and shore applications during the past two decades. Several sizes and shapes have been used to satisfy various space limitations, and various feed techniques have been used. Antennas of this type with a lower frequency limit of 2 mHz are quite practical for shore station use, whereas limitations on physical size confine their use to frequencies higher than 6 mHz on shipboard.

Figure 64 shows the configuration of a 2 to 32 mHz discage antenna. This antenna combines the characteristics of the conical monopole (top-loaded) and the discone into one compact structure. The cage, like the previously described conical monopole, is fed at its base ring or hoop. The discone, which is formed of the top disc and the upper portion of the cage, is fed by a coaxial line which runs up through the central mast or tower. The outer conductor of the coaxial line connects to the upper portion of the cage, and the inner conductor connects to the disc. Thus there are two separate lines, one for the high band feed to the disc, and the other for the low band feed to the cage. The disc provides toploading under certain conditions of termination of the discone's feedpoint. This top-loading is desirable at the cage's lower frequencies for reasons of impedance, but it is not desirable at the cage's higher frequencies from the standpoint of the vertical pattern shape. Therefore the terminating conditions of the discone's feedpoint must be controlled, or switched. More on this point later.

There are several basic considerations which must be observed in the design of this antenna. First, the overall height is chosen on the basis of the lowest operating frequency, with the toploading taken into consideration. The transition frequency between the operating bands of the discone and the cage is very important. It determines the maximum diameter of the cage and the overall dimensions of the discone. Electrical height of the overall structure at the transition frequency must be less than ¾ wavelength to avoid undue vertical pattern lobing at the cage's high frequency limit. For the discone, the flare angle of the cone should be about 60°, the diameter of the disc should be at least 0.7 times the maximum diameter of the cone, and the disc-to-cone spacing should be about 0.3 times the diameter of the apex of the

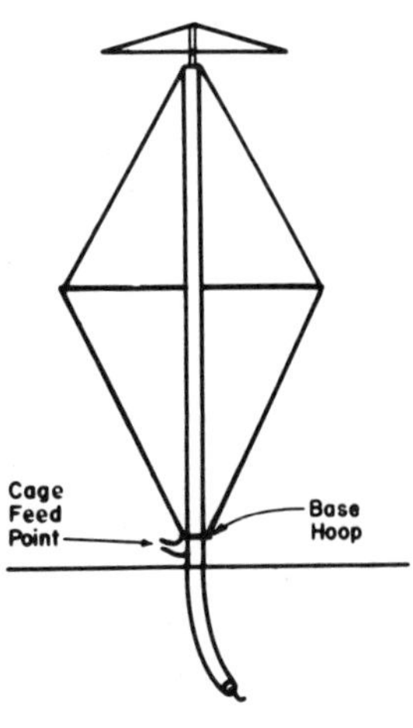

Fig. 64—Configuration of the discage antenna, a combination of the discone and conical monopole antennas. The coaxial feedline runs up through the mast; the inner conductor connects to the top disc and the outer conductor connects to the cage at the top. The second feedline connects to the cage feed point at the base hoop.

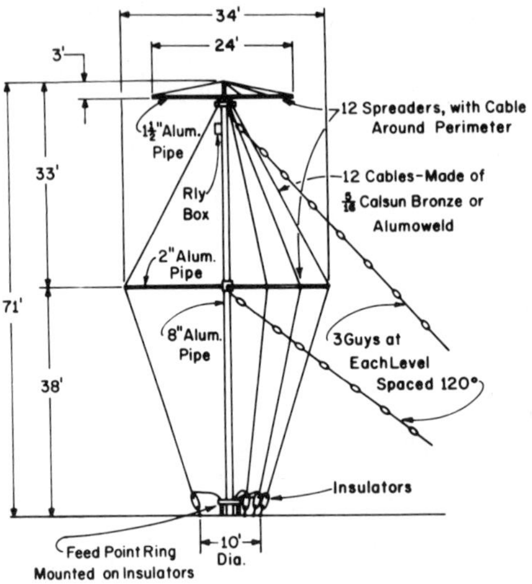

Fig. 65—Dimensions of a 2-32 mHz discage antenna. The relay box mounted at the top of the 8″ aluminum mast houses a relay which shorts out the disc when the cage is being fed.

cone. The slant height of the discone's cone should be approximately ¼ wavelength at the discone's lowest operating frequency. It so happens that all these parameters fall together to make the choice of about 8 mHz a good one for the transition frequency, for a 2 to 30 mHz discage. This is near the geometric mean of the limits of the frequency range.

As to the number and size of conductors, the simulation of the solid disc and cage improves as the number and diameter of the conductors is increased. There is, however, a practical limit to this. Experience has shown that a cage made of 12 conductors performs very well. A top disc simulated by 12 spreaders and interconnecting wires is used. Figure 65 shows this configuration. Aluminum pipe, or a small aluminum tower, can be used as the center support. As with the conical monopole, the structure lends itself to being guyed by use of some of the cables as portions of guys, isolated from the remainder of the guys by strain insulators. The disc can be mounted to the top of the mast or tower as shown in fig. 66. A circular plate is used to secure the upper ends of the cone cables. Another circular plate rests on this, supported by four standoff insulators. The disc spreaders are bolted to the plate. A small supporting rod is mounted in the center of the upper plate, to hold the cables which support the far ends of the spreaders. A waterproof relay box is mounted on the side of the mast. Power for the relay is run up through the mast, and, inasmuch as the mast is grounded at its base, the power line can be merely bypassed to ground with a pair of 0.1 mf 600 volt capacitors at the base of the mast.

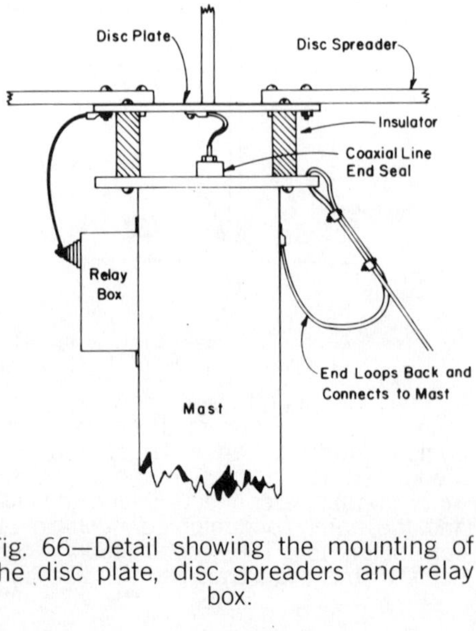

Fig. 66—Detail showing the mounting of the disc plate, disc spreaders and relay box.

It was mentioned previously that the terminating conditions of the discone must be either controlled or switched in order to make use of the top-loading effect of the disc at the cage's low frequency limit. In shipboard applications[44] we often feed this antenna with two or more transmitters, through multicouplers, with several transmitters operating in the high band and several in the low band, all operating simultaneously. In order to do this, and to prevent unwanted interactions between high and low band equipments, it is necessary to use a high pass filter in the coaxial line to the discone, and a low pass filter in the line to the cage. The high pass filter is so designed as to properly terminate the discone so that the disc will act as top-loading at the low frequency limit of the cage. The high pass filter is located up on the mast near the feed point of the discone. However, for amateur applications, where we are not operating transmitters simultaneously, this termination can be switched by means of a relay, mounted on the mast as shown. The relay shorts the disc to the cage, when operating in the low frequency portion of the cage's band.

The dimensions shown in fig. 65 are for a low limit of 2 mHz. Throughout the 2 to 8

[44]Wilson, K. D., "Communications, A Limiting Factor in Naval Warfare," *Naval Engineers' Journal,* Feb. 1963, p. 51.

mHz band and the 8 to 32 mHz band the s.w.r. will be less than 2.5 to 1, referred to 50 ohms. If it desired to operate this antenna down to 1.8 mHz, its size can be increased by a factor of 10%, or one can be content to tolerate an s.w.r. of 3 to 1 in this region. If one wishes to start at a low limit of 3.5 mHz, the size can be scaled down by a factor of 0.57, thus giving an antenna which would cover 80, 40, 20, 15, 10 and 6 meters. If one is only interested in 7 mHz and up the size can be reduced by a factor of 0.286, and 40, 20, 15, 10, 6 and 2½ meters could be covered. The height of a 3.5 mHz version would be 39.7 feet and that of a 7 mHz version would be only 19.85 feet.

In the naval service, for shipboard use, we do one other little trick with this antenna, and that is to mount a small u.h.f. unipole antenna on top of it, utilizing the disc as the ground plane for the u.h.f. antenna. One way of carrying the u.h.f. feed line across the h.f. discone feed point would be to coil up the u.h.f. feed line in the form of a choke coil which would have little effect on the h.f. discone feed point. Such a choke could be used to terminate this feed point, instead of using the shorting relay. The choke would thus act as a small loading coil between the cage and the top disc, at the low frequency limit of the cage. This is shown schematically in fig. 67.

As for vertical patterns, fig. 68 shows those for the cage, with the discone feed point shorted, for 2.5, 4.0, 5.5 and 7.5 mHz. In fig. 69 we see the vertical patterns of the discone portion, with the cage feed terminated. While the latter show multiple lobes, this antenna is considered to be quite suitable for omnidirectional medium and long distance communications.

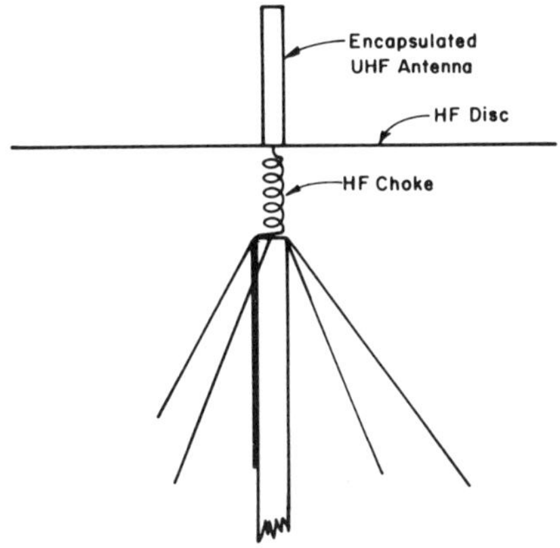

Fig. 67—Configuration of a method of u.h.f. feed across the h.f. discone. The u.h.f. coaxial line is wound into an inductance, the h.f. choke. The u.h.f. line then runs down through the h.f. antenna mast.

The discage antenna would be ideal for someone who is limited as to space and who wants one antenna to cover a number of frequencies. While the Mark IV DX Antenna is ideal for amateur band use only, it is not suitable for MARS use, for example, without the addition of more relays and matching networks in the tuning unit. A discage antenna, on the other hand, enables one to operate on *any* frequency without the use of tuning networks or switching. Power at 2 to 8 mHz is fed in through one coaxial line, and the other line is used for 8 to 32 mHz operation. It is reasonable in size, even for 2 mHz. ■

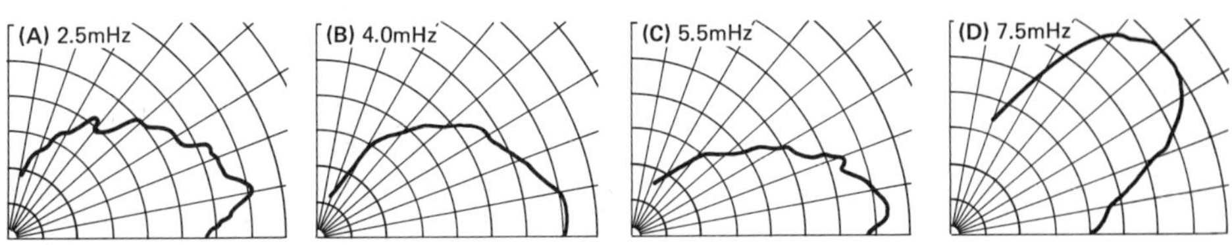

Fig. 68 — Vertical patterns of the cage with the discone feed point shorted.

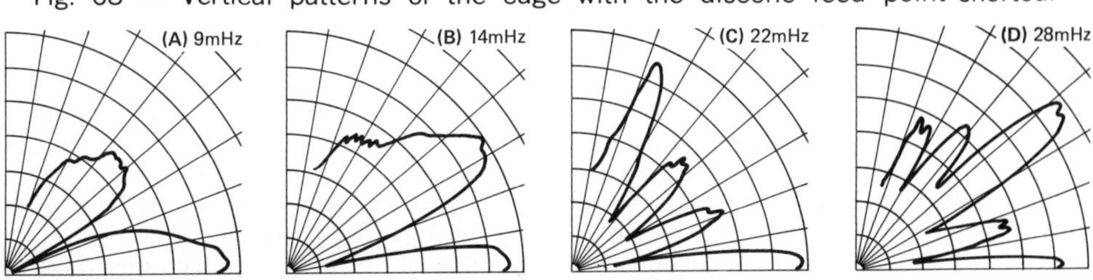

Fig. 69—Vertical patterns of the discone portion with the cage feed terminated.

BROADBAND VERTICAL ANTENNA CONFIGURATIONS

Several additional broadband vertical antennas are described in this chapter. These antenna types can very easily be adapted to amateur use as shown by the author.

IN Chapter Seven it was stated that conical structures make ideal broadband radiators, and their use, both as inverted cones and discones, being fed against a flat planar surface, was shown. This matter can be carried further by stating that the structure does not have to be conical but can be pyramidal in shape. This configuration lends itself to the joining of two structures together to form a folded unipole. Such a configuration is shown in fig. 70 in perspective view. The two pyramids rest on their apexes, with a junction at their bases. One apex is grounded, and the other is fed. A ground system is required.

The Folded Unipole

This antenna was used on the "Voice of America" ship *USCGC Courier,* which for a period of about ten years was moored in the eastern Mediterranean Sea as a short wave and medium wave relay broadcast station. This facility has since been decommissioned and replaced by shore based equipment and antennas, but these broadband structures were very interesting and unusual. There were two of them on the ship. One was for a frequency range of 6 through 11 mHz, and the other was used for 9 through 17 mHz. The structures were made of steel angle and bar stock, assembled around a strong central steel pipe, and completely welded to provide not only mechanical strength but also thorough electrical bonding of all pieces. This was necessary because of the maritime environment which in a short time would have produced many "rusty bolt" connections in an unwelded structure, and also because of the power involved (35 kw 100% modulated).

Figure 70 shows the dimensions of this antenna for several frequency ranges of interest to amateurs. It is capable of being used over a frequency range of 3.5 to 1 with an s.w.r. of less than 1.5 to 1 at the feed point, referred to the nominal design impedance of 175 ohms which was desired for these antennas. For 50 ohm feed it was necessary to use a simple L network as shown in fig. 71 with shunt capacity and series inductance. Even with the network in use the s.w.r. will still be good over quite a wide frequency range, such as the entire 3.5 to 4.0 mHz band for example, without returning.

The 175 ohm design impedance results from the use of a rather acute angle at the apex of each pyramid. The apex angle of the structures shown in fig. 70 is about 30°. If the apex angle is increased to about 60° (the same value as used in the discone antenna of fig. 53), this impedance will come down to a more reasonable

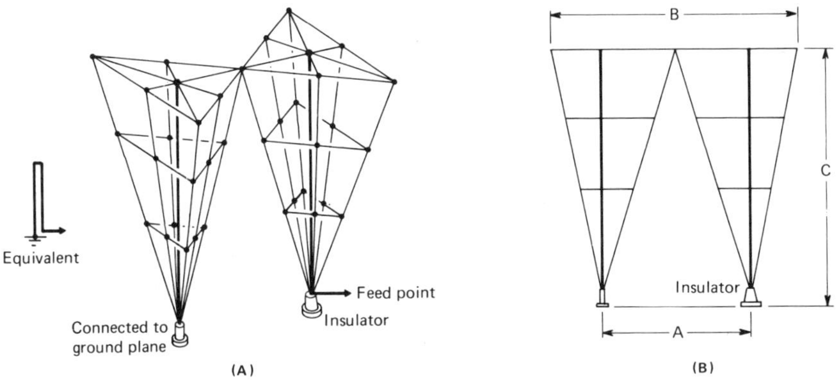

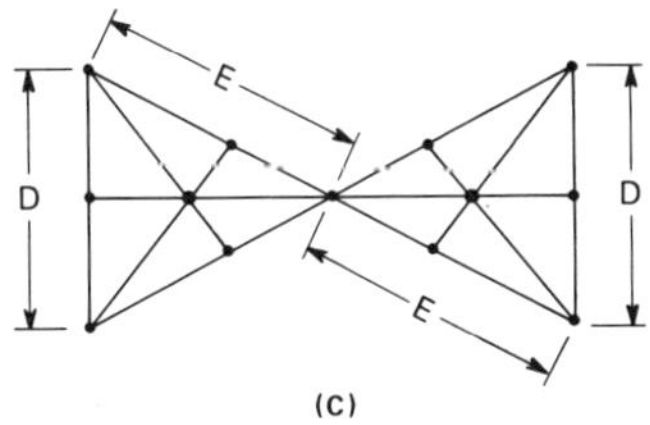

Dimension	3.0-10.5 mHz	7.0-24.5 mHz
A	27′	11.6′
B	45′	19.3′
C	45′	19.3′
D	20′	9.6′
E	25′	10.7′

Fig. 70 (A) — Perspective view of a triangular folded unipole and its equivalent circuit. (B) and (C) — Side and top views of the triangular folded unipole and the critical dimensions for two frequency ranges of interest to amateurs.

value of 50 to 75 ohms, and the excursions of the reactance versus frequency curve will be less pronounced.[45, 46] One might assume this to be the case, from what has been said previously about broadband structures, for as the apex angle is reduced in value to 0° we have a linear folded monopole whose input impedance will have extreme variations versus frequency. With the 60° apex angle, the dimensions of fig. 72 will apply. It will be noted that the horizontal area required for the antenna has increased over that needed for the 30° case. This was another factor affecting the shipboard installation, where space was limited.

The amateur could build an antenna of this type using pieces of aluminum tubing flattened on their ends and bolted together. The entire structure would be quite light weight, and would require only four lateral guys and two longitudinal guys.

The Folded Fan

Another arrangement which is somewhat similar to this is a folded fan type of antenna which is being used on shipboard. An entire portion of a ship's superstructure such as a stack, fire control director or radar tower, *etc.*, is driven as a part of a folded fan type of unipole, as shown in fig. 73. The amateur who has available a tall metal structure such as a chimney or tower can apply this type of feed to it and obtain quite good broadband results. In fact, if the tower is a thin one, it can be "fattened" by the use of a second fan as shown in fig. 74. For matching 50 ohm line to this type of feed, "cut and try" will be required, for I cannot give any absolute values of reactances required. However, if the total height of the folded fan is less than 1/4 wave, the impedance should be quite reasonable (several hundred ohms or less) and easy to match with a simple T or L network.

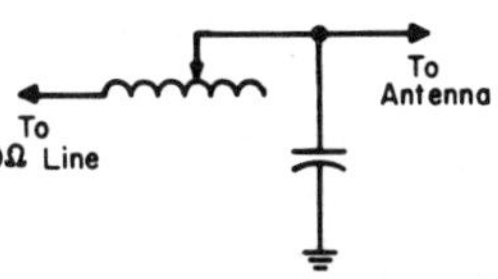

Fig. 71 — Circuit of the L network used to match the triangular folded unipole's 175 ohm feed impedance to a 50 ohm line.

[45] Jasik, Henry, *"Antenna Engineering Handbook," McGraw-Hill, p. 3-11, 3-12.*

[46] Brown, G. H. and Woodward, O. M., "Experimentally Determined Radiation Characteristics of Conical and Triangular Antennas," *RCA Review*, Dec. 1952.

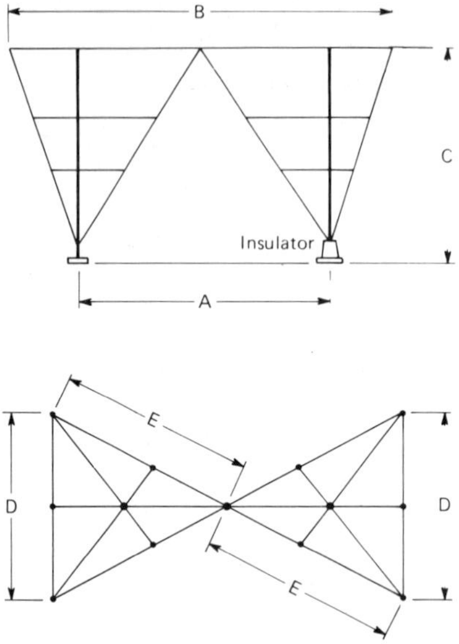

Dimension	3.0-10.5 mHz	7.0-24.5 mHz
A	56′	24′
B	90′	38.6′
C	45′	19.3′
D	51′	21.8′
E	51′	21.8′

Fig. 72 — Dimensions of a triangular folded unipole with an apex angle of 60°. This angle, instead of the 30° shown in fig. 70, will provide a feed impedance of 50 to 75 ohms.

The Log Periodic

The types of antennas described above are nominally non-directional, with perhaps 1 to 3 db variation in the horizontal pattern polar plot. Let us now consider a *directional broadband* radiator, the vertical log-periodic antenna. Where in the previous cases we were using simple linear structures and making them broadband by lowering the effective L/D ratio, the log-periodic is of a different breed. This type of structure is composed of thin linear elements, and its geometry is defined so that the radiation pattern and the impedance repeat periodically with the logarithm of frequency. Some basic design theory of log-periodic antennas is in order here, applying to all types and not only to vertical ones, in order that their operation can be understood.

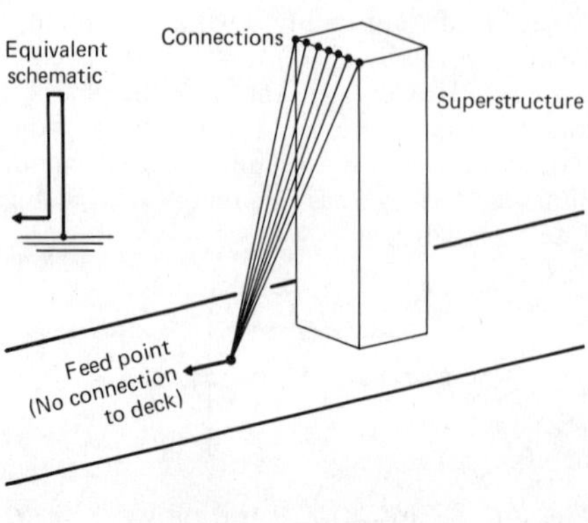

Fig. 73 — Folded fan feed of a portion of a ship's superstructure.

Figure 75 shows a basic type of log-periodic antenna, used here merely to show the relationships of the various dimensions. (Never mind how it is to be fed at this time.) This antenna is a trapezoidal tooth structure made of wire. One of the basic parameters is the "design ratio," which is called τ. Also related to it is σ, which is equal to $\sqrt{\tau}$.
The following equations apply:

$$\tau = \frac{R_{n+1}}{R_n} = \frac{r_{n+1}}{r_n}$$

$$\sigma = \frac{r_n}{R_n} = \frac{r_{n+1}}{R_{n+1}} = \sqrt{\tau}$$

The design ratio must not be less than about 0.7 for good gain and low back lobes.

Another important parameter is the element spacing-to-length ratio.

$$\text{Spacing } S_n = R_n - R_{n+1} = R_n(1-\tau)$$

and the length of the n-th element is:

$$L_n = R \tan\frac{\alpha}{2}$$

and the element spacing-to-length ratio, therefore, is:

$$\frac{S_n}{L_n} = \frac{1-\tau}{\tan\frac{\alpha}{2}}$$

Experimental work has shown that this ratio should not exceed a value of 1.30 for proper operation of the antenna. Thus for a given angle α, there is a minimum value of τ.

$$\tau_{min} = 1 - 1.3 \tan\frac{\alpha}{2}$$

The frequency range of an antenna of this type is determined by the length of the longest and shortest elements, which are 1/2 wavelength at

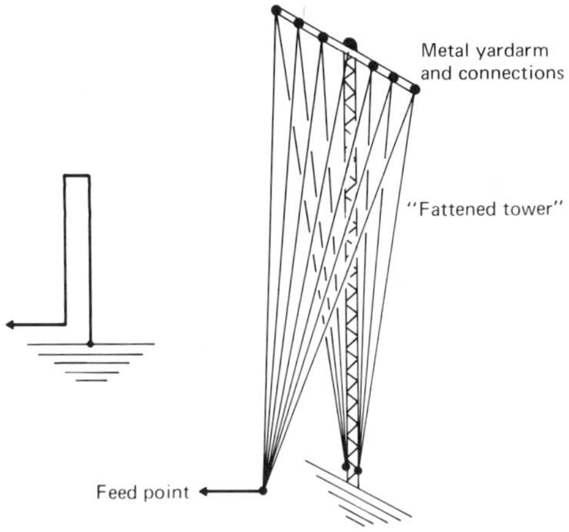

Fig. 74 — Broadband folded fan on a tower.

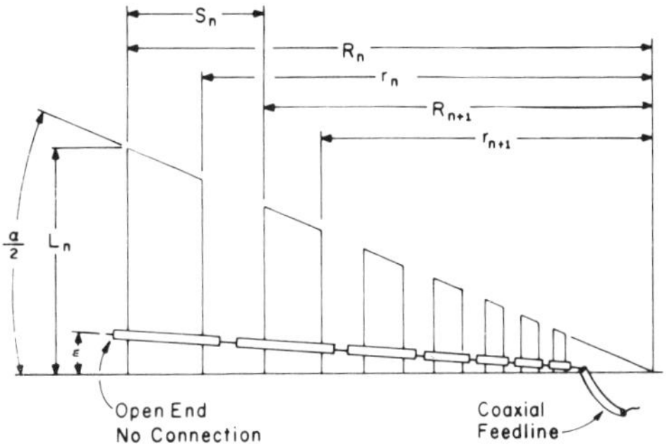

Fig. 76 — A vertical log-periodic antenna.

the lowest and highest frequencies, respectively. For all practical purposes, the detail with which the small end can be built determines the high frequency limit.

It can be seen from the above that there are a number of interrelated parameters in the design of a log-periodic antenna. As the antenna is made shorter for a given frequency range, the number of elements decreases, the design ratio decreases, and the front-to-back ratio and gain decrease. On the other hand, a longer antenna is more expensive to construct, and requires more real estate.

The configuration shown in fig. 75 is a balanced one. There are several ways of constructing and feeding this one, and antennas of this type are usually used in the horizontal mode, with a rotary mounting for directional operation, with the direction of the main lobe off the small end of the antenna. Since we are dealing with vertical antennas in this series, consider fig. 76, which is one half of the above antenna, with the lower half replaced by a ground system. A practical configuration which will give good results is shown. The front-to-back ratio at the lower frequencies will be about 14 db, while at the higher frequencies it will be about 20 db, with the value of $\alpha/2$ shown.

This antenna is fed by means of a capacitive coupling arrangement. The outer conductor of the coaxial feedline is broken between the trapezoidal teeth, as shown, and each piece is connected to one of the teeth forming one plate of a capacitor. The inner conductor is the other part of the capacitor, and is common to all teeth. The bottom end of each tooth is connected to the ground system. Angle ε is about 3°. If the s.w.r. is to be optimized, the angle ε should be varied somewhat, while input impedance is plotted on a Smith Chart, over the antenna's frequency range. With 50 ohm line, it is possible to obtain an s.w.r. below 3 to 1 over the frequency range of such an antenna. A frequency range of 10 or 15 to 1 is easily possible, and the horizontal and vertical patterns hold shape remarkably well over the entire range. This type of antenna is an excellent one

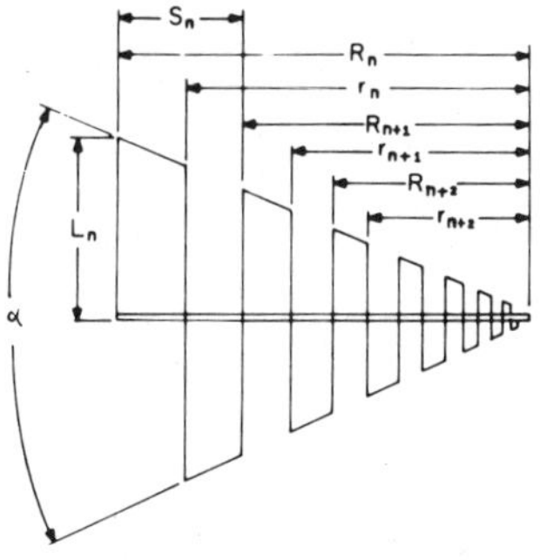

$$\tau = \frac{R_{n+1}}{R_n} = \frac{r_{n+1}}{r_n}$$

$$\sigma = \frac{r_n}{R_n} = \frac{r_{n+1}}{R_{n+1}}$$

$$\sigma = \sqrt{\tau}$$

Fig. 75 — Basic log-periodic antenna structure and some of the design formulas discussed in the text.

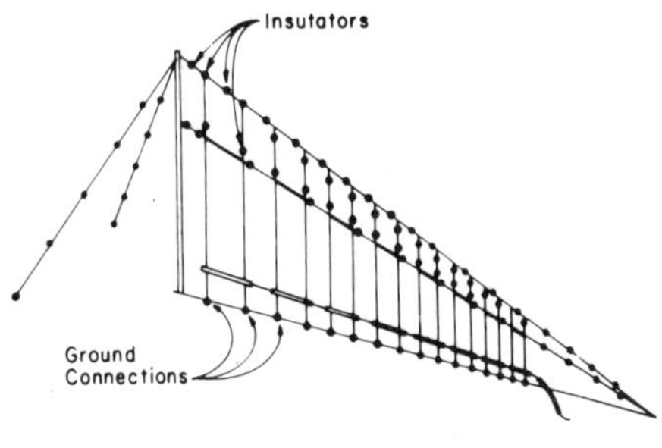

Fig. 77 — Sketch of a wire type log periodic constructed by suspending the teeth from a catenary strung from a tower to an anchor block.

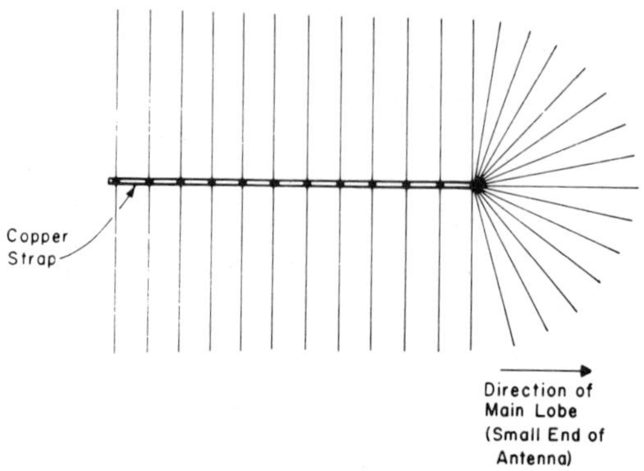

Fig. 78 — Radial system for a vertical log periodic. The radials should be at least a quarter wavelength at the low frequency limit of the antenna and all joined by a 2" wide copper strap.

for those who wish to cover a wide range of frequencies, with transmission and reception in a particular direction. Again I repeat that the longer the antenna, the greater the number of teeth, and the larger the design ratio τ, the higher will be the gain and the front-to-back ratio. For the one shown in fig. 76 I have not shown any dimensions. Dimensions will depend on the frequency range to be covered. Obviously an antenna for a range of 2 to 30 mHz will be longer and have more teeth than one for a range of 7 to 30 mHz. The parameters will be the same, however.

The actual construction is quite simple. The teeth, made of wire, can be supported on a catenary suspended from one tall pole or tower to an anchor block in the earth, as shown in the sketch of fig. 77. The ground system can be a buried wire mat of the type shown in fig. 78, with a center buss of two inch wide copper

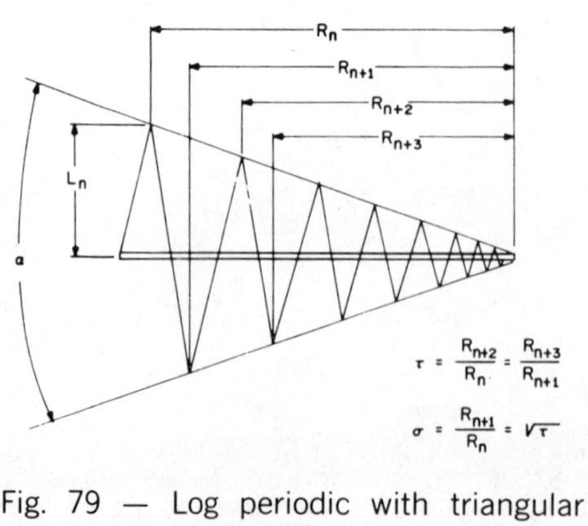

Fig. 79 — Log periodic with triangular teeth.

strap, to which the bottom ends of the teeth must be bonded.

The teeth of the log-periodic do not have to be a trapezoidal shape. They can also be triangular in shape, as shown in fig. 79. In this case, the following equations apply:

$$\tau = \frac{R_{n+2}}{R_n} \text{ and}$$

$$\sigma = \frac{R_{n+1}}{R_n} = \sqrt{\tau}$$

As in the case of the trapezoidal teeth, the design ratio τ must not be less than about 0.7 for good gain and low back lobes.

$$\textit{Spacing } S_n = R_n - R_{n+2} = R_n(1 - \tau),$$

and the length of the n-th element is:

$$L_n = R_n \tan\frac{\alpha}{2},$$

and the element spacing-to-length ratio; therefore, is:

$$\frac{S_n}{L_n} = \frac{1 - \tau}{\tan\frac{\alpha}{2}}$$

As in the previous case, this should not exceed a value of 1.30 for proper operation of the antenna. And as in the previous case,

$$\tau_{min} = 1 - 1.3 \tan\frac{\alpha}{2}$$

For the case of the triangular tooth structure working against ground, one must actually design the structure of fig. 79, and then cover up one half of it with a plain sheet of paper, leaving only the configuration shown in Fig. 80. The teeth of this one can be supported by a catenary, with single point suspension for each tooth. This triangular tooth model is perhaps somewhat easier to rig than the one with trapezoidal teeth. It can be fed in the same manner as shown in fig. 76, with coaxial line with broken outer conductor acting as sections of a capacitor.

It is possible to build an array of antennas of this type by using one common center supporting pole for the catenaries, and extending the catenary cables for two, three or four of these antennas out in the desired directions of transmissions. Each one would have its own feed cable, and thus the station operator would have a choice of directions by switching antennas. In a case of this kind, the ground systems should be connected together where they intersect.

In addition to the references listed[47, 48], there are many others in the technical literature.

[47] DuHamel, R. H. and Ore, F. R., "Logarithmically Periodic Antenna Designs," 1958 *IRE National Convention Record,* Part 1, p. 139-151.

[48] DuHamel, R. H. and Berry, D. G., "Logarithmically Periodic Antenna Arrays," 1958 *IRE WESCON Convention Record,* Part 1, p. 161-174.

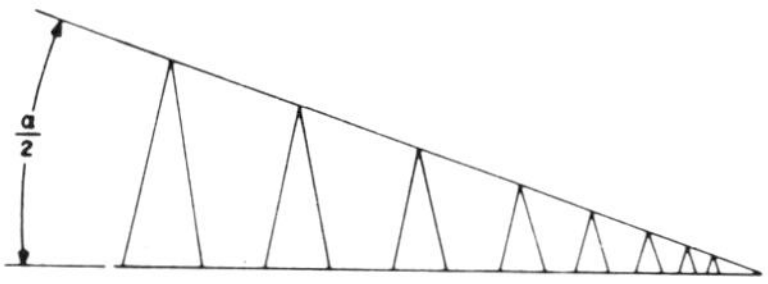

Fig. 80 — Log periodic with triangular teeth for operation against ground.

Jasik[45] is an excellent source of reference listings on antennas of all types, including the log-periodics.

DESIGNING A SPECIFIC DIRECTIONAL VERTICAL ANTENNA INCLUDING FEED SYSTEM

Chapter Five gave general information on directional arrays. In this chapter the author discusses the design of a specific array and its feed system. This array is easily adaptable to multi-element switchable configurations for changing direction of transmission.

SINCE the publication of the "Vertical Antennas" series in CQ in 1968-1969, there has been considerable reader correspondence concerning the detailed design of directional arrays and their feed systems. Most of the inquiries have concerned the two-element array with 90° spacing and 90° phasing, and ways of switching three or four elements in pairs in order to use this two-element array for transmission in several selectable directions. Examples of this are three elements in an equilateral triangle, or four elements in a square, with 90° sides, with means of swtiching to any pair of them at one time.

A	B	C	D	E	F	G	H
ϕ	Cos A	SxB	C + ψ	Cos D	1 + E	$\sqrt{F}$	K x G
0°	1.000	90.0	180.0	-1.00	0.	0.	0.
10°	.985	88.7	178.7	- .99	.01	.1	13.8
20°	.940	84.6	174.6	- .99	.01	.1	13.8
30°	.866	77.9	167.9	- .98	.02	.14	19.3
40°	.766	68.9	158.9	- .93	.07	.26	35.9
50°	.643	57.9	147.9	- .85	.15	.39	53.8
60°	.500	45.0	135.0	- .71	.29	.54	74.5
70°	.342	30.8	120.8	- .51	.49	.70	96.5
80°	.174	15.7	105.7	- .27	.73	.85	117.3
90°	0	0	90.0	- 0	1.00	1.00	138.0
100°	-.174	-15.7	74.3	.27	1.27	1.12	154.6
110°	-.342	-30.8	59.2	.51	1.51	1.23	169.7
120°	-.500	-45.0	45.0	.71	1.71	1.31	180.7
130°	-.643	-57.9	32.1	.85	1.85	1.36	187.7
140°	-.766	-68.9	21.1	.93	1.93	1.39	191.8
150°	-.866	-77.9	12.1	.98	1.98	1.40	194.6
160°	-.940	-84.6	5.6	.99	1.99	1.41	195.6
170°	-.985	-88.7	1.3	.99	1.99	1.41	195.6
180°	-1.000	-90.0	0	1.00	2.00	1.41	195.6

Table 1 — Data for the computation of antenna radiation pattern.

Outlined below are the complete details of the design of a two-element array with 90° spacing and 90° phasing, and the design of its feed system and antenna matching units. This configuration will give only 3 db gain over the radiation from a single element, in spite of some claims of 6 db made by others. *It is impossible to obtain 6 db gain from two elements in any configuration.* Three elements are required. Maximum in-line gain from two elements is 4.74 db, and this is with approximately 30° spacing and 165° phasing. Readers are referred to references 27, 28 and 29 for complete details of directional designs and patterns, and gains obtainable.

Pattern Computation

The first step in the design of any array is the pattern computation. In these days of computers, an engineer can compute the complete horizontal and vertical patterns of any array quite rapidly and accurately. He can even design an array to produce a pattern that will fit a given set of radiation limits at specific azimuths and vertical angles. But such computations are a rather laborious process if one does not have a digital computer handy. It requires use of a slide rule and trigonometric tables, and a quick ability to visualize angles of more than 90° and their sines and cosines. For pattern computation for ground wave only, and where the two tower currents are equal, the equation:

$$E = K\sqrt{1 + \cos(S\cos\phi + \psi)}$$ is

an easy one to use, where S is the spacing in degrees, ϕ is the azimuth angle, ψ is the phase angle, and K is a constant which depends on the r.m.s. of the mathematical plot.

For the array in question, S is 90°, ψ is 90°, ϕ is taken every 10° from 0° through 180°, and the computations are tabulated as shown here. For simplicity in labelling the columns, each is given a capital letter designation in Table I.

Columns A through G can be computed directly from the equation and the given values of S, ϕ and ψ. The next step is the determination of K, in order to produce column H which is a tabulation of unattenuated field intensity at one mile in millivolts per meter, for 500 watts radiated power using two quarter wave elements. The first thing to be done is to find the r.m.s. of the pattern. This is done by adding up Column F, multiplying it by 2 (to get the whole pattern from 0° through 350°), then subtracting one set of values for 0° and 180° (so that they are not added in twice), dividing by 36, then taking the square root of what remains, thus:

Sum of column F = 19.00
× 2
38.00
− 0.00
38.00
− 2.00
36.00

$$\frac{36.00}{36} = 1.00$$

$\sqrt{1.00}$ = 1.00 r.m.s. of pattern.

1.00 = 138 mv/m for /4 elements.

K = 138

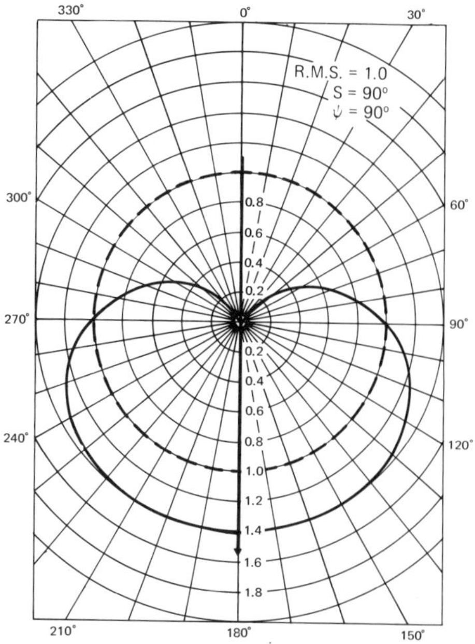

Fig. 81 — Polar pattern for the design problem illustrated in the text.

The pattern can be plotted in mv/m, in which case Column H is plotted, or it can be plotted normalized to its r.m.s. When one is

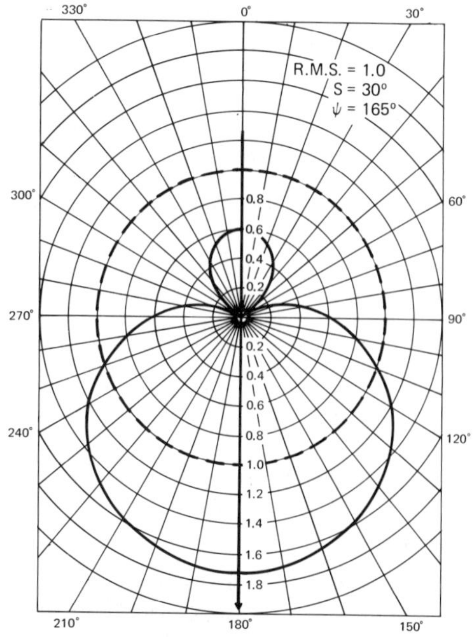

Fig. 82 — Polar pattern for a vertical array with a spacing of 30º and a phasing of 135º

A	B	C	D	E	F	G	H	I	J
ϕ	Cos A	.985B	SxC	D+ψ	Cos E	1+F	$\sqrt{G}$	.965H	KxI

Table II — Headings needed for tabulation of data for vertical pattern computation.

discussing array theory and design, the plot is usually normalized to an r.m.s. of 1.00 so that the plot shows the gain or loss at any particular azimuth angle ϕ. The plot of the above computations is shown on this basis in fig. 81, which is the same as fig. 32. It will be seen that the voltage (field intensity) gain at 180° is 1.41. The power gain is therefore 2, and this equals 3 db.

In the event that you are a bit confused by the mathematics above, let's say that in the case of an in-line array the pattern will be the same on both sides of the line of towers. Hence the tabulation stopped at 180°. There is no need to carry it on to 360°, for the values will be identical to those already tabulated. However, in computing the r.m.s., one must include 36 items, and only 36. Hence the doubling of the sum of Column *F*, and the subtraction of one set of 0° and 180° items which would otherwise appear twice in the summation. It is purely happenstance that the sum of Column *F* comes out as it does, and that the r.m.s. of the pattern is 1.00, in the above case. It does so because of the 90° phasing and 90° spacing. It would not do so with other values. For example, if I were to tabulate the computations for the case of S = 30° and ψ = 165° (maximum in-line gain of 4.74 db), the sum of Column *F* would be 1.871, and when the r.m.s. is computed it would be 0.308. For an r.m.s. of 138 mv/m, *K* would be 448. Figure 82 is a plot of this pattern, normalized to an r.m.s. of 1.00.

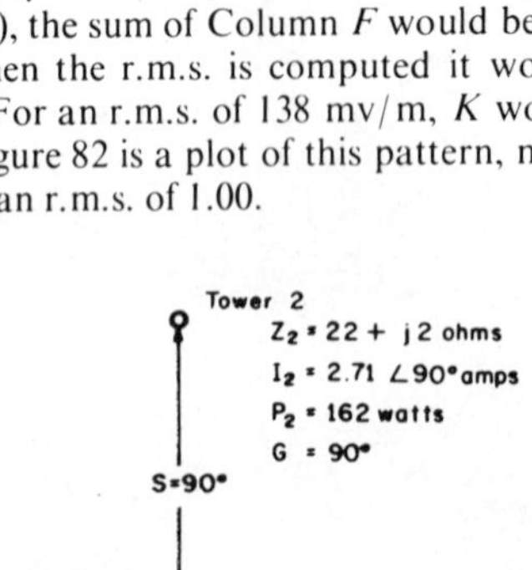

Fig. 83 — Array configuration of the design example discussed in the text.

Vertical Pattern

In commercial practice we often need to know the vertical pattern as well. This can be computed in a like manner, by inserting the vertical angle θ into the formula, as shown here:

$$E = Kf(\theta)\sqrt{1 + \cos(S\cos\phi\cos\theta + \psi)}$$

The vertical shape factor $f(\theta)$ was discussed in Chapter Five, and it depends on the tower height *G* in degrees. For the case of quarter wave towers used here, *$f(\theta)$ is 0.965 when θ* is 10°. Cos θ is 0.985 when θ is 10°. The table of computations would thus look like Table II as a start. There is no point in finishing it as it would not accomplish very much except to use up space. A similar table would be required for each value of θ, every 10 degrees. In broadcast consulting work, this is required by the FCC for directional arrays and this is done by computer for rapidity and accuracy. Let me summarize by saying that for in-line arrays, the vertical pattern of the main lobe in the line of towers is generally a bit more flat than that from one tower alone, the vertical pattern broadside to the line of towers is the shape of that of one tower alone, and the vertical pattern at other angles has to be computed to be known accurately. But in our case, let's not worry about the exact vertical pattern of this array, and get on with the design of the system.

Element Self-Impedance

Now that we have the array parameters, how do we feed it? The first step is a determination of the self-impedance of the elements (which are called towers). The array designer has to start somewhere, and that place is with published plots of *R* and *jX* versus tower height in wavelengths. Let's assume that this array is to be designed for 3.9 mHz. A tower height of 60 feet would be about right for a quarter wavelength. An effective diameter of 7 inches would be reasonable. L/D would therefore be about 100. Referring back to Chapter three it will be found that K_a for this value of *L/D* is 599. Or K_a can be computed from the equation:

$$K_a = 120\left(\log_e \frac{21}{a} - 1\right)$$

Referring to figure 18, for a tower height of a quarter wave and K_a of 600, *R* = 34 ohms and $X = +j25$. The next step is the computation of the effect of the mutual impedance between the towers, and the actual operating impedances of each tower. Referring to Fig. 37,

Fig. 84 — Power dividing circuit for the array design of fig. 76. Phase shifts and impedances of various points are noted. Line 1 is 90º longer than line 2 and the relative phase shifts are 90º and 0º.

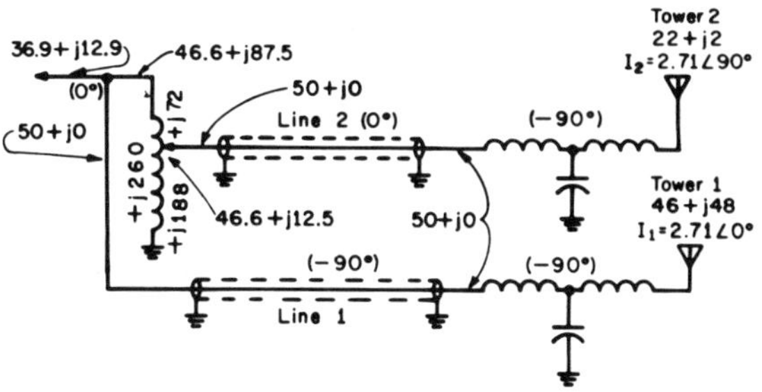

we find that the mutual impedance between towers 1 and 2 equals 23—j12 ohms, for the 90° spacing of two quarter wave elements.

The equations for the operating impedances of the two towers when connected in the array area:

$$Z_1 = Z_{11} + \frac{I_2}{I_1} Z_{21} \text{ and}$$

$$Z_2 = Z_{22} + \frac{I_2}{I_1} Z_{12}$$

where Z_{11} is the self-impedance of tower 1 by itself,

Z_{22} is the self-impedance of tower 2 by itself,
I_1 is the base current (loop current) of tower 1,
I_2 is the base current (loop current) of tower 2,
Z_1 is the actual operating impedance of tower 1 in the array, and
Z_2 is the actual operating impedance of tower 2 in the array.

We just found Z_{11} and Z_{22} from fig. 18 and they are 34 + j25. *Since we don't know the currents yet, and we are only interested in their ratio and phase angle at this time, let* I_1 = $1.0\underline{/0^\circ}$ and $I_2 = 1.0\underline{/90^\circ}$. Mutual impedance $Z_{21} = Z_{12} = 23 - j12$.

Solving for Z_1:

$$Z_1 = (34 + j25) + \left(\frac{1.0\,/90^\circ}{1.0\,/0^\circ}\right)(23 \times j12)$$
$$= (34 + j25) + (1.0\,\underline{/90^\circ})\,(26\,\diagdown\overline{27.5^\circ})$$
$$= (34 + j25) + (26\,\underline{/62.5^\circ})$$
$$= (34 + j25) + (12 + j23)$$
$$= 46 + j48 \text{ ohms}$$

Solving for Z_2:

$$Z_2 = (34 + j25) + \left(\frac{1.0\,/0^\circ}{1.0\,/90^\circ}\right)(23 - j12)$$
$$= (34 + j25) + (1.0\,\diagdown\overline{90^\circ})\,(26\,\diagdown\overline{27.50})$$
$$= (34 + j25) + (26\,\diagdown\overline{117.5^\circ})$$
$$= (34 + j25) + (-12 - j23)$$
$$= 22 + j2 \text{ ohms}$$

Note the difference between these values and the self-impedance of the towers! The mutual impedance and the currents flowing cause quite an effect!

Now that we have the actual operating impedances of the two towers, we can find the currents.

$$P_1 = I_1^2 R_1$$
$$P_2 = I_2^2 R_2$$
$$P_1 + P_2 = 500 \text{ watts}$$
$$I_1 = I_2$$
$$I_1^2 R_1 + I_1^2 R_2 = 500$$
$$I_1^2 (R_1 + R_2) = 500$$
$$I_1^2 (46 + 22) = 500$$
$$I_1 = 2.71 \text{ amps.}$$

As a check:

$$P_1 = (2.71)^2\, 46 = 338$$
$$P_2 = (2.71)^2\, 22 = 162$$
$$P_{Total} = 162 + 338$$
$$= 500 \text{ watts}$$

The array configuration is shown in fig. 83. This brings up one point to be remembered, which is that in the case of an in-line array, the main lobe is always in the direction of the tower with the lagging phase (in this case I_1 lags I_2 by 90°).

Power Divider

Now that we have the tower operating impedances, it is possible to design the power dividing circuit and the antenna tuning units. Again referring to Chapter Five, we shall use the Ohms Law network. Since we have only

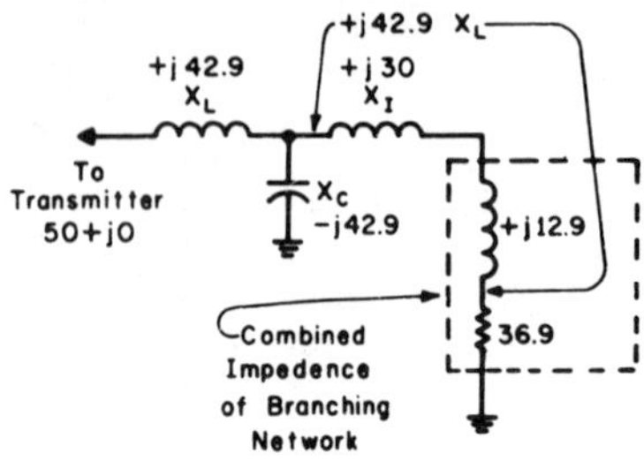

Fig. 85 — Input T network.

two towers, and one of them is taking 2.1 times the power taken by the other, it is quite simple. We can feed tower 1 directly from the branching point, and tap down on a power divider coil for the feed to tower 2. See fig. 84 for the complete network.

How do we arrive at the various values of reactance shown? First, since the array is to work on 3.9 mHz we assume a 10 microhenry power divider coil will be used. This is a reasonable value. Since the power ratio between the towers is 2:1, tower 2 feed is tapped down on the coil at a 72.5% point. The total coil reactance at 3.9 mHz is $+j72$ ohms above the tap. Considering the feedline to be a pure 50 ohms resistive load, the parallel combination of 50 and $+j188$ is computed:

$$\frac{(50 \angle 0^\circ)(188 \angle 90^\circ)}{50 + j188} = \frac{9400 \angle 90^\circ}{195 \angle 75^\circ} = 48.2 \angle 15^\circ = 46.6 + j12.5$$

Add the upper portion of the coil to this, to find the total impedance of the tower 2 feed which is presented to the branching point:

$$46.6 + j12.5 + j72 = 46.6 + j87.5 \text{ ohms.}$$

Next, add in parallel the 50 ohm pure resistance of tower 1 line and the above combined value for tower 2 feed:

$$\frac{(46.6 + j87.5)(50 \angle 0^\circ)}{50 + 46.6 + j87.5} = \frac{(99.1 \angle 67^\circ)(50 \angle 0^\circ)}{96.6 + j87.5} = \frac{4960 \angle 62^\circ}{130 \angle 42.2^\circ} = 38.1 \angle 19.8^\circ = 36.9 + j12.9 \text{ ohms}$$

This is the parallel combination of the tower 1 and tower 2 feeds at the common point. As may be seen, it is quite a long way from being a pure 50+j0 ohms which a transmitter would like to see. Since we have only two towers, and this impedance value is so close to 50 ohms, we can dispense with the tuned tank arrangement shown in fig. 38 and transform this 36.9 + j12.9 ohms to 50 + j0 with a simple T network. (Another reason for the tuned tank in broadcast applications is that it enables the operator to maintain the common input impedance to licensed value quite easily. In this application we are not concerned with this.)

Another very helpful trick which one soon learns in this business is the usefulness of a 90° network,. It is so easy to compute and to use, where the 90° phase shift is of no consequence! In a 90° network, where R_1 and R_2 are the input and output terminating resistances, the values of X_C *and* X_L in the network are determined very easily from the formula:

$$X = \sqrt{R_1 R_2}$$

The T network for the input of this branching system is shown in fig. 85. In this case:

$$X = \sqrt{36.9 \times 50} = 42.9$$

X = —j 42.9 or 950 mmf at 3.9 mHz.
X = +j42.9 or 01.85 mh at 3.9 mHz.
X = +j42.9 — j12.9 = +j30 or 0.14 mh at 3.9 mHz.

Tuning Units

Now we must turn our attention to the antenna tuning units. Working our way outward from the branching circuit, tower line 1 must have 90° more electrical length in it than tower line 2. If it does not, a T phase shift network would be necsessary. A 90° T for 50 ohms would consist of series X_L = +j50 elements, and shunt X_C = —j50 elements, which would have to be variable to take care of strays. It is much easier to make the lines 90° different in length, and take care of any strays by slight adjustment of the antenna tuning units (ATUs).

As in the case of the input T, we again make use of the easy design of the 90° T in the ATUs. For tower 1, the ATU is shown in fig. 86.

$$X = \sqrt{46 \times 50}$$

$$X = 47.9$$

$X = -j47.9$ or 850 mmf at 3.9 mHz
$X = +j47.9$ or 1.85 mh at 3.9 mHz.
$N = +j47.9 - j48 = -j0.1$

In this case it would be necessary to put a series capacitor in, because X_1 is so close to zero, it could go a bit either way in actual adjustment. Series 50 ohms is 800 mmf at 3.9 mHz. X_1 should then be made about $+j75$, adjustable.

ATU number 2 is done in the same way. It is shown in fig. 87.

$$X = \sqrt{22 \times 50} = 33.2$$

$X = -j33.2$ or 1230 mmf at 3.9 mHz
$X = +j33.2$ or 1.35 mh at 3.9 mHz.
$N = +j33.2 - j2 = +j31.2$ or
1.27 mh at 3.9 mHz.

This, then, is how we design a directional antenna system with its phasing and branching networks and antennas tuning units. In actual practice the values of reactance required are given about a 50% increase in tolerance to take care of differences between measured and computed antenna impedances and mutual impedances. The first step is the design shown here. The next step is construction of the array and its ground system, and actual r.f. bridge measurement of tower and mutual impedances. These should be fairly close to the predicted values. Using the *measured* values, the professional engineer then runs through the phasing and branching network calculations *again*, to determine as closely as possible the values to which all reactances should be set. Coils are variable, either with movable taps or of the rotary variety. Fixed capacitors are used, in series with adjustable coils, to obtain the $-jX_C$ values required. (Capacitive reactance is made too large, and a portion is cancelled out by a variable coil.) The engineer then uses the r.f. bridge to set all components to their final *computed* values. Power is then applied (low power at first), and tower currents and phases are read. Field intensity readings are taken in the nulls in the pattern. Final adjustments are made, sometimes over a period of many nights (midnight to 6 A.M.), to bring the array in to its design values and the nulls down to FCC required values of field intensity. The r.f. bridge used in such work is either a General Radio 916A, 916AL, or 1606, whose accuracies are excellent.

For the amateur, the proof of successful operation would be the reading of tower currents, and a check of the forward gain, or the depth and azimuth of the designed nulls. This

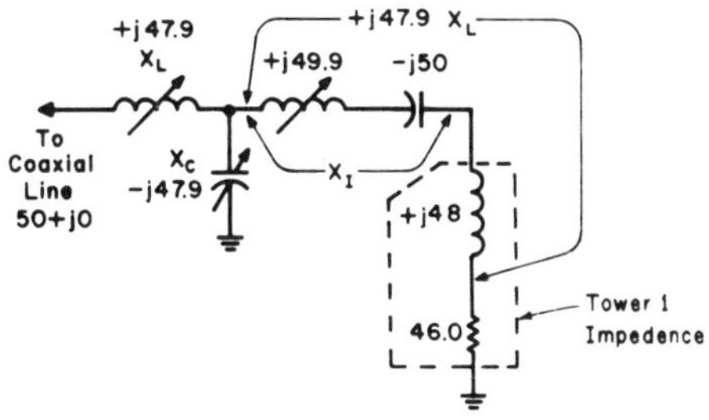

Fig. 86 — Antenna tuning unit (ATU) for tower 1.

can be done more easily and with less hard work than in the broadcast case where tolerances must be very close. If a good check on the operation of the above array is desired, it can be obtained by first operating one tower alone, with the other one floating or detuned, and measuring field intensity around it at various azimuths. If it is in the clear, with no other objects to absorb or distort the field, a fairly circular pattern should be obtained. With a known power input, the radiation efficiency can be obtained by plotting field intensity versus distance as described in Chapter Two. The array can then be connected, and with the same input power, its field intensity should be measured *at the same points* as with the single tower. This pattern can then be plotted. The ratio of the directional to non-directional field intensity at at a particular point or along a particular radial is the gain (or loss) of the array at that azimuth. Unless there are unpredictable things such as houses, power line, other antennas, *etc.*, distorting the field, the plot should be close to the original design plot. In no case will the two-element array shown, with 90° spacing and 90° phasing, produce more than approximately 3 db in the main lobe.

One additional check which is useful is that the r.m.s. value of the directional pattern should approach very closely the value of the field of the circular non-directional pattern,

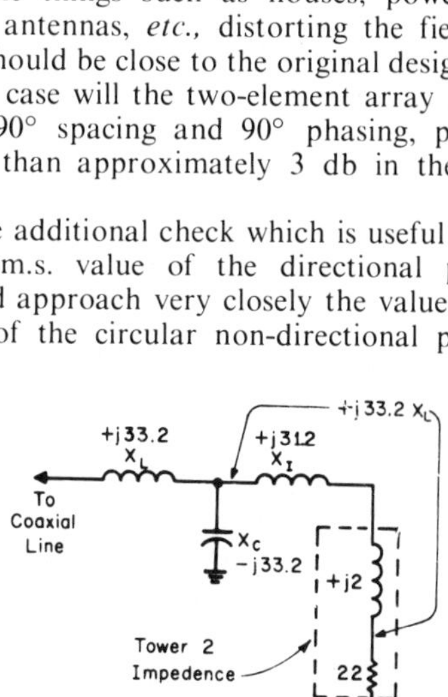

Fig. 87 — Antenna tuning unit for tower 2.

with the same input power. It may be slightly lower due to added losses in the second tower and coupling systems.

Arrays of this configuration have been built without ATUs or branching circuits, feeding power to the towers by merely connecting coaxial lines to them and paralleling them, with one 90° longer than the other. These do work after a fashion, but the adjustment can hardly be optimum without some means of assuring proper power division and equal tower currents. ■

10

USING AN EQUILATERAL TRIANGLE CONFIGURATION FOR A SIMPLE DIRECTIONAL ARRAY

The author discusses a very simple directional array which can be built in an equilateral triangle configuration. Using any two elements at a time world wide coverage can be obtained at a gain of 5.2 db.

As previously stated in Chapter Five, the maximum gain obtainable with two vertical elements in any configuration is 5.2 db. This occurs with an element spacing of 210°, and with equal in-phase currents. Readers should refer to references 27, 28 and 29 for details of this and other directional configurations, and gains obtainable. I should also reiterate here that when talking about gain of an array of equal height elements, the reference is the non-directional field which would be radiated from one single element of the same height when the total power fed to the array is fed to it. The field from a single element radiating 500 watts would be 138 mv/m at one mile, for example. This field, minus any losses due to the directional array's phasing and branching system (usually less than 2% for a two element array), will be equal to the r.m.s. of the directional pattern. Thus, when plotting a directional pattern, we can use the ratio of the pattern's field in any direction to its r.m.s. as the pattern's voltage gain in that particular direction. Knowing the voltage gain, we can find the gain in db easily.

Equilateral Triangle

The case of two elements spaced 210° with equal in-phase currents lends itself very nicely to an equilateral triangle configuration, with the elements fed two at a time by means of a switching arrangement. This case is also a very

A	B	C	D	E	F	G	H	J
ϕ	Cos A	S x B	C + ψ	Cos D	1+E	$\sqrt{F}$	K x G	1.29xG
0°	1.000	210.0	210.0	-.866	.134	.366	65.1	.472
10°	.985	206.7	206.7	-.894	.106	.326	48.2	.421
20°	.940	197.3	197.3	-.954	.046	.214	38.1	.276
30°	.866	182.0	182.0	-.999	.001	.031	5.5	.040
40°	.766	160.9	160.9	-.945	.055	.234	41.7	.302
50°	.643	135.0	135.0	-.707	.293	.542	96.4	.699
60°	.500	105.0	105.0	-.259	.741	.861	153.2	1.110
70°	.342	71.8	71.8	+.313	1.313	1.143	203.5	1.474
80°	.174	36.5	36.5	+.804	1.804	1.342	239.0	1.731
90°	0	0	0	1.000	2.000	1.414	252.0	1.825
100°	-.174	-36.5	-36.5	+.804	1.804	1.342	239.0	1.731
110°	-.342	-71.8	-71.8	+.313	1.313	1.143	203.5	1.474
120°	-.500	-105.0	-105.0	-.259	.741	.861	153.2	1.110
130°	-.643	-135.0	-135.0	-.707	.293	.542	96.4	.699
140°	-.766	-160.9	-160.9	-.945	.055	.234	41.7	.302
150°	-.866	-182.0	-182.0	-.999	.001	.031	5.5	.040
160°	-.940	-197.3	-197.3	-.954	.046	.214	38.1	.276
170°	-.985	-206.7	-206.7	-.894	.106	.326	48.2	.421
180°	-1.000	-210.0	-210.0	-.866	.134	.366	65.1	.472

Table III — Data for the computation of the antenna radiation pattern shown in fig. 88

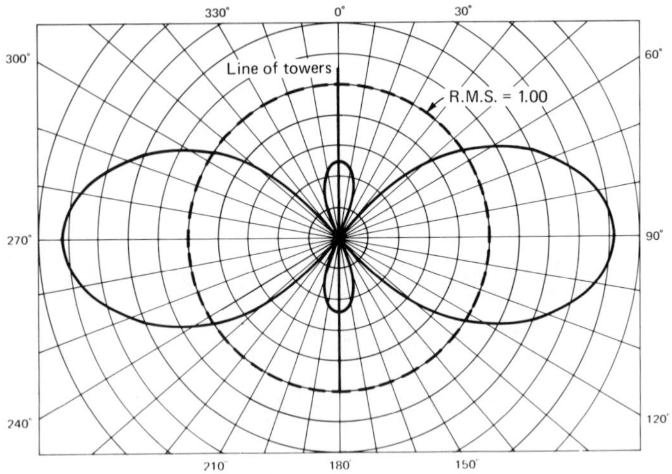

Fig. 88 — Antenna radiation pattern, plotted from Column J of Table III for the two elements 210º apart with equal, in phase, currents.

easy one to compute. Inasmuch as the currents in the equal height elements are equal and in phase, the elements' operating impedances will be equal, they will be fed equal amounts of power, and their antenna tuning units will be identical. This makes the switching arrangement very simple, for the triangular configuration.

First, let's run through the pattern computation for this one, as done in Chapter Nine, inasmuch as we do not have a computer. For ground wave pattern computation, and where the two tower currents are equal, the equation used is:

$$E = K\sqrt{1 = cos\,(S\cos\phi + \psi)}$$

where S = spacing in degrees
ϕ = azimuth angle
ψ = phase angle
K = constant depending n r.m.s. of the pattern's mathematical plot.

For the array being discussed here, S is 210°, ψ is 0°, and ϕ is taken every 10° from 0° through 180°. The tabulation of computations is shown in Table III. Each column is given a capital letter designation for simplicity in labelling the columns in the heading of the table. The computation follows through very logically, step by step, column by column.

The pattern will be symmetical about the line of towers, so it is necessary to compute only from 0° through 180° the values for columns A through G. The r.m.s. of the pattern is determined by adding column F, multiplying it by 2 to get the whole pattern from 0° through 350°, then subtracting one set of values for 0° and 180° (so that they are not included twice in the summation), dividing by 36, then taking the square root of what remains:

$$\begin{array}{rr} \text{Sum of column } F = & 10.986 \\ & \times\quad 2 \\ \hline & 21.972 \\ & -.134 \\ \hline & 21.838 \\ & -.134 \\ \hline & 21.704 \end{array}$$

$$\frac{21.704}{36} = 0.603$$

$$\sqrt{0.63} = 0.776 \text{ r.m.s.}$$

$$K = \frac{138}{0.776} = 178$$

Column H shows the value of the radiated field at the various values of ϕ, in mv/m at one mile. Column J shows the pattern normalized to a value of 1.000 for its r.m.s., for purposes of determining voltage gain. The values in column J are plotted in fig. 88. The gain in the direction of the maximum lobes (broadside to line of towers) is 1.825, which is a power gain of 3.33 or 5.23 db. There are two minor lobes, with two pairs of nulls at 31°, 149°, 211°, and 329°. I might state here that if the spacing were 180° instead of 210°, with equal in-phase currents, the two minor lobes would not be present. There would be only two nulls, at 0° and 180°, but the gain at 90° and 270° would be reduced to 4.56 db. Any reader who wishes to run through the pattern computation for this latter case is welcome to do so, and find out this answer for himself.

If one wishes to compute the vertical pattern in the case of this 210° spacing, 0° phasing array, it may be done by use of the expanded formula:

$$E = Kf(\theta)\sqrt{1 + \cos(S\cos\phi\cos\theta + \psi)}$$

Factor $f(\theta)$ was discussed in Chapter Five, and it depends on the tower height G in degrees. θ is the vertical angle for which the value of radiation is to be computed.

Feed Design

With the array parameters decided, we must now determine how to feed it. The first step is the determination of the self-impedance of each tower. Assuming a design frequency of 3.9 mHz, a tower height of 60 feet is about right for a quarter wavelength. Assuming an effective diameter of 7 inches, the L/D of the tower would be about 100. Referring back to Chapter Three it will be found that K_a for this value of L/D is 599. Or K_a can be computed from the equation:

$$Ka = 120\left(\log_e \frac{21}{a} - 1\right)$$

From fig. 18, for a tower height of a quarter wave and a K_a of 599, R = 34 ohms and X = + j25 ohms. Next comes the computation of the effects of mutual impedance between the towers, and the actual operating impedance of each tower. Again referring to fig. 37, we see that the mutual impedance between the towers spaced 210° is —9 —j11 ohms. Yes, —9 ohms! Don't let the minus sign in front of the R term frighten you. It is perfectly "legal."

The equations for finding the actual operating impedances of each tower are:

$$Z_1 = Z_{11} + \frac{I_2}{I_1} Z_{21} \text{ and,}$$

$$Z_2 = Z_{22} + \frac{I_1}{I_2} Z_{12}$$

where Z_{11} is the self-impedance of tower 1 by itself,

Z_{22} is the self-impedance of tower 2 by itself,

I_1 is the base current (loop current) of tower 1,

I_2 is the base current (loop current) of tower 2,

Z_1 is the actual operating impedance of tower 1 in the array, and

Z_2 is the actual operating impedance of tower 2 in the array.

This is the same set of equations used in Chapter Nine.

Having found Z_{11} and Z_{22} from fig. 18, we know them to be 34 + j25 ohms. The mutual impedances Z_{21} and Z_{12} equal —9 —j11 ohms. We do not know the magnitude of I and I_2 yet, but they are equal and in phase, and can be represented as $I_2 = I_1 = 1.0\angle 0°$.

Solving for Z_1:

$$Z_1 = (34 + j25) + \left(\frac{1.0\angle 0°}{1.0\angle 0°}\right)(-9 - j11)$$

$$Z_1 = (34 + j25) + (1)(-9 - j11)$$

$$Z_1 = 25 + j14 \text{ ohms}$$

Solving for Z_2:

$$Z_2 = (34 + j25) + \left(\frac{1.0\angle 0°}{1.0\angle 0°}\right)(-9 - j11)$$

$$Z_2 = (34 + j25) + (1)(-9 - j11)$$

$$Z_2 = 25 + j14 \text{ ohms}$$

Note that we get the same result. This means that the operating impedance of the two towers are identical. Note, however, how different they are from the tower self-impedances. The effect of the mutual impedance and the currents flowing is quite pronounced.

Tower Currents

Now that we know the actual operating impedance we can find the tower currents:

$$P_1 = I_1^2 R_1$$

$$P_2 = I_2^2 R_2 \text{ and}$$

$$P_1 + P_2 = 500 \text{ watts}$$

$I_1 = I_2$ *and, in this case,*

$R_1 = R_2$

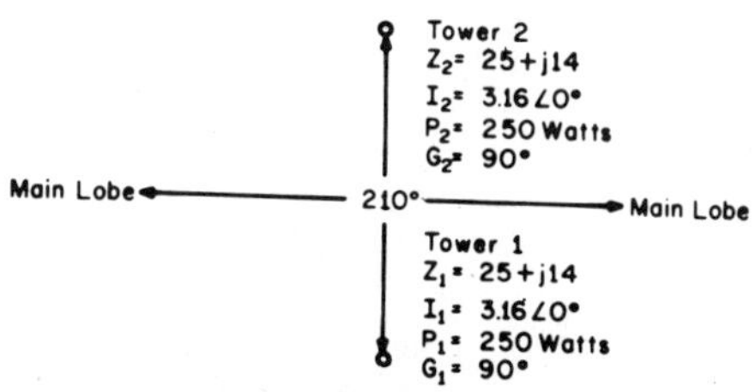

Fig. 89 — Array configuration discussed in text.

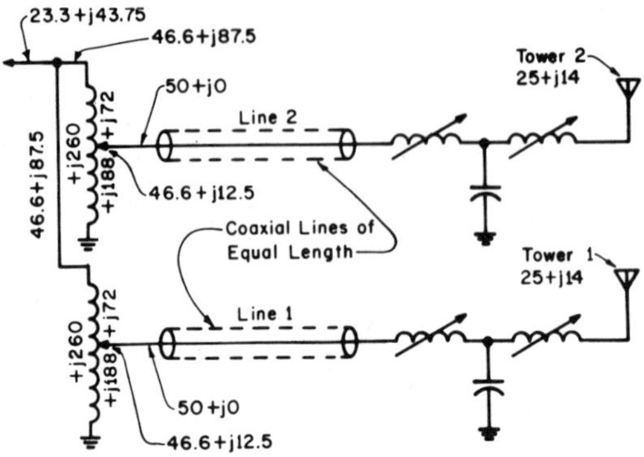

Fig. 90 — Branching circuit and the various impedances and reactances.

Therefor:

$$2\,(I^2_1 R_1) = 500$$
$$2\,(I_1^2 \times 25) = 500$$
$$I^2 = 10$$
$$I_1 = 3.16 \text{ amps} = I_2$$

As a check:

$$P_1 = (3.16)^2\, 25 = 250$$
$$P_2 = (3.16)^2\, 25 = 250$$
$$P = 250 + 250 = 500 \text{ watts}$$

The array configuration is shown in fig. 89.

ATU'S

Now that we know the tower operating impedances, it is possible to design the power dividing circuit and the antenna tuning units. It would be possible to do without any tuning units and simply connect 50 ohm line to the towers, and parallel the two equal length lines at the feed point. However, quite a large s.w.r. will result, and this is undesirable. The circuit of fig. 90 is recommended. An Ohm's Law network is used to give exact control over power division. Since the array is to work on 3.9 mHz, 10 microhenry power divider coils are assumed. Each of these has a reactance of $+j260$ ohms at 3.9 mHz. The taps are at equal distances above ground, and are placed at the $+j188$ ohm points, to make the computation simple. This leaves $+j72$ ohms above each tap. Considering each feedline to be pure 50 ohms resistance load, the paralel combination of 50 and $+j188$ is computed:

$$\frac{(50\,\angle 0^\circ)(188\,\angle 90^\circ)}{50 + j188} = \frac{9400\,\angle 90^\circ}{195\,\angle 75^\circ}$$

$$48.2\,\angle 15^\circ = 46.6 + j12.5$$

Adding the upper portion of the coil to this, the total impedance of each tower's feed which is presented to the common branching point is:

$$46.6 + j12.5 + j72 = 46 + j87.5$$

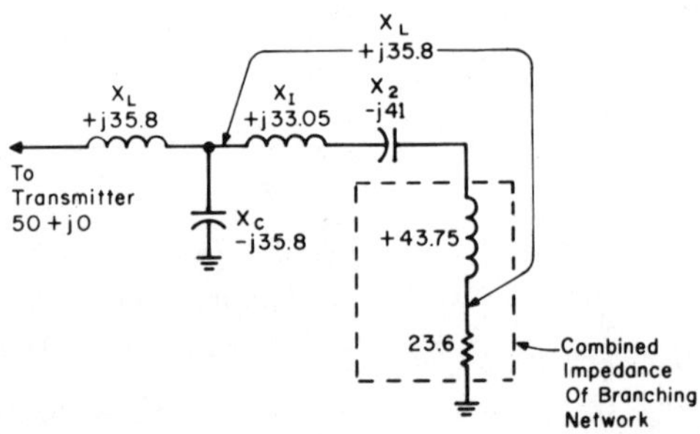

Fig. 92 — ATU employs a 90° network for matching.

Inasmuch as the two tower feeds are equal adding them in parallel can be accomplished by dividing one of them by 2:

$$\frac{46.6 + j87.5}{2} = 23.3 + j43.75$$

These several impedances and reactances are shown at appropriate places in fig. 90.

A matching network will be required to transform this value of $23.6 + j43.75$ to $50 + j0$ ohms. Again, as in Chapter Nine, we use our old friend the 90° network, so easy to compute. It is shown in fig. 91. The values of Xc and X_L are easily computed from the equation:

$$x = \sqrt{R_1 R_2}$$

In this case:

$$x = \sqrt{23.3 \times 50}$$

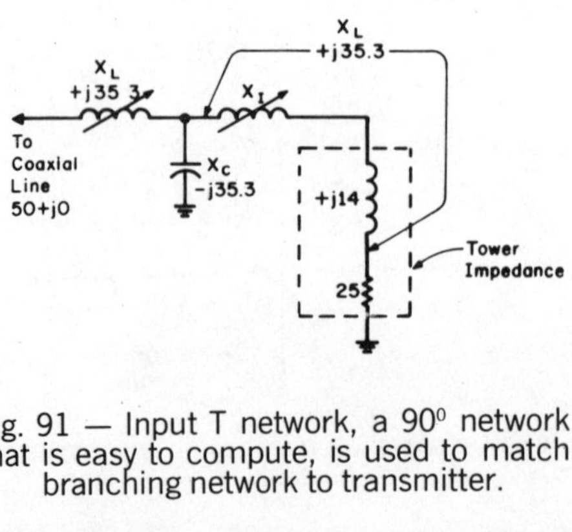

Fig. 91 — Input T network, a 90° network that is easy to compute, is used to match branching network to transmitter.

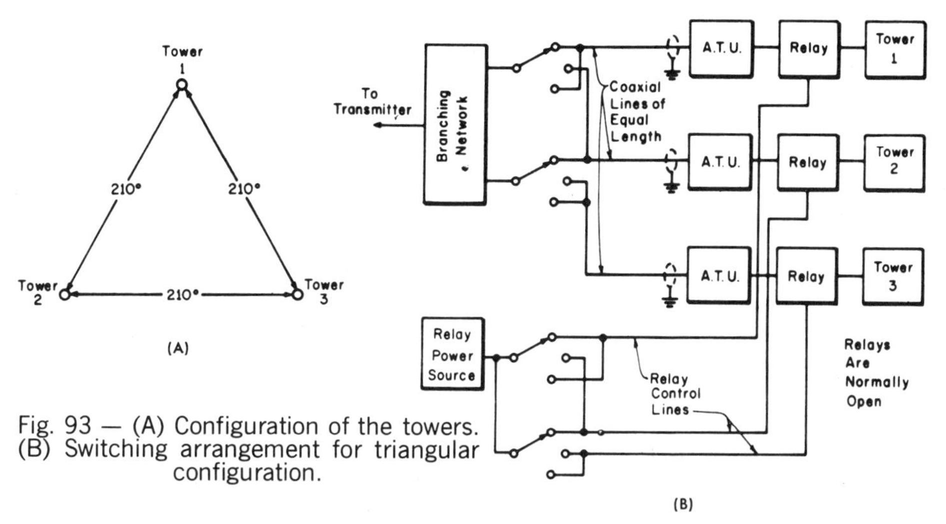

Fig. 93 — (A) Configuration of the towers. (B) Switching arrangement for triangular configuration.

$$= 35.8$$
$$x = -j35.8 \quad \text{or} \quad 1120 \text{ mmf at } 3.9 \text{ mHz.}$$
$$x = +j35.8 \quad \text{or} \quad 1.45\mu\text{h at } 3.9 \text{ mHz.}$$

Inasmuch as X_L is less than the reactance of the input to the branching network, it is most expedient to insert some negative reactance X_C in series and to cancel some of it out with an adjustable positive X_L. Why not simply insert a negative reactance equal to the difference between $+j43.75$ and $+j35.8$ ohms? This would be equal to about $-j8$ ohms, which would be 5000 mmf at 3.9 mHz. It would be a bit impractical to make this variable. It is much easier to make the inductance variable. In this case a 1000 mmf fixed capacitor is chosen for X_C, with a reactance of $-j41$ ohms. Thus X_1 can be found as follows:

$$x_1 = +j35.8 - (+j43.75 - j41)$$
$$x_1 = +j33.05 \quad \text{or} \quad 1.35\mu\text{h at } 3.9 \text{ mHz.}$$

Now, turning our attention to the antenna tuning units and noting that the tower operating impedances are identical, the ATUs will be identical. (Figs. 90 and 92.) Again using our old friend the 90° T network, so easy to handle,

$$x = \sqrt{25 \times 50}$$
$$= 35.3$$

$$x_{C_1} = -j35.3 \quad \text{or} \quad 1106 \text{ mmf at } 3.9 \text{ mHz.}$$
$$x_C = +j35.3 \quad \text{or} \quad 1.47\mu\text{h at } 3.9 \text{ mHz.}$$
$$x_C = +j21.3 \quad \text{or} \quad 0.886\mu\text{h at } 3.9 \text{ mHz.}$$

It must be emphasized that this is the *computed* design. In actual field practice the professional engineer will measure the tower self-impedances and the mutual impedance with his General Radio r.f. bridge, and will then run through the phasing and branching network calculations again using the actual *measured* values. The results should be fairly close to the computed values. In actual practice, the equipment designer will make all reactances with a 50% increase in tolerance to allow for the differences which may occur between computed and measured values. To obtain odd values of $-jXc$, it is standard practice to use a fixed capacitor in series with a variable coil.

Three Element Array

Now that we have the array and feed system designed, it is quite simple to use it in an equilateral triangle configuration, with three elements, driving any pair of them at a time. Such a system is shown in schematic form in fig. 93. A three position rotary switch is needed, with provision for open-circuiting the unused tower in each position. This is most easily done by means of a relay in each tower lead.

The triangle can be oriented so that the bidirectional beam from each pair of elements falls on a desired azimuth, for global coverage, as shown in fig. 94. This is an example of the directions which can be obtained from a west coast site. A similar map can be used for any other site in the United States. Azimuthal maps such as this often appear in the front pages of the *Radio Amateur Call Book,* and they may be obtained also from the U.S. Navy

Fig. 94 — Directions of the main lobes as shown on an azimuthal projection centered on San Diego Line A-A is the direction of the main lobes from towers 1 and 2, B-B from towers 2 and 3 and C-C from towers 1 and 3. The tower orientation for this pattern is discussed in the test.

Oceanographic Office, Washington, D.C., and the U.S. Coast and Geodetic Survey, Washington, D.C.

In fig. 94, Line *A-A* is the direction of the main lobes from towers 1 and 2, Line *B-B* is the direction of the main lobes from towers 2 and 3, and Line *C-C* is the direction of the main lobes from towers 1 and 3. To produce the main lobes oriented on the azimuths shown, the triangular configuration of towers shown in fig. 93(A) is oriented as follows:

Line of towers 2 and 1—bearing 10°
Line of towers 2 and 3—bearing 70°
Line of towers 1 and 3—bearing 130° ■

THE OPERATION AND DESIGN OF THE FOLDED UNIPOLE ANTENNA

A folded unipole antenna less than a quarter wave long can be used to transform the low input resistance of a short vertical antenna to a resistance which is more reasonable to match and drive. The author discusses the operation and design of the folded unipole antenna in this chapter.

VERY low input resistance is often present in antennas. This is due to the antenna being very short electrically or to the mutual impedance present from another nearby antenna. The coupling network required to match a low resistance (of the order of one or two ohms, or less) to a 50 ohm line is not difficult to compute, but it may include some rather impractical values of reactance. A short series fed antenna looks capacitive, and its capacitive reactance is usually quite high; the shorter the antenna the higher the reactance. The series coil necessary to resonate it (called a "helix" in low frequency or very low frequency work) usually contributes considerable loss in itself, and it is therefore desirable to get rid of it if possible and use some other method of matching and feeding power to the antenna. The Type UG and the NORD antennas described in Chapter Four are designs which accomplish this. It will be noted that each of these uses a folded unipole type of feed. Folded antennas as a class or type provide a means of securing the desired impedance transformation within the antenna itself, thus simplifying the required matching network.

Folded Dipole

It is likely that all readers are familiar with the folded dipole. Some may be using one for transmitting, but it is much more probable that one is being used for TV reception, as the folded dipole is the "driven element" of most such v.h.f. Yagi arrays. The reason for its use in that application is the very one mentioned previously, to present a more favorable im-

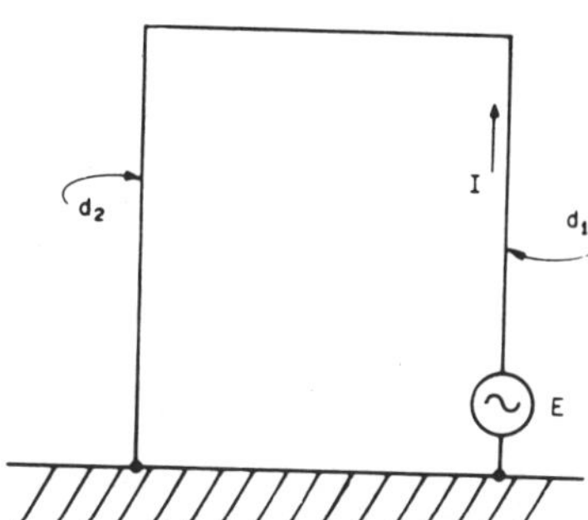

Fig. 95 — Basic circuit of the folded unipole antenna.

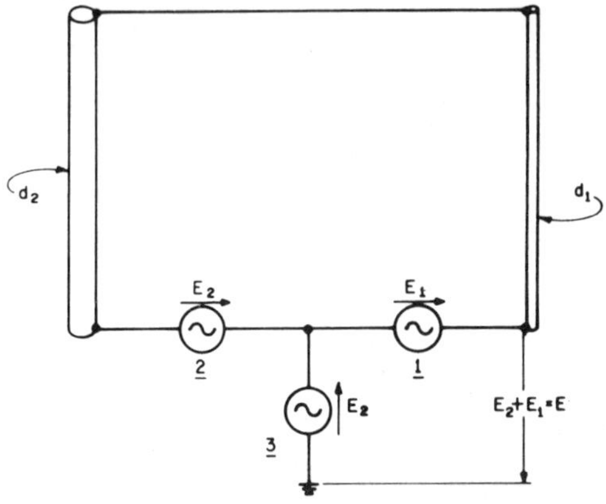

Fig. 96 — Equivalent circuit of the folded unipole that can be used for analysis.

pedance match to the 300 ohm transmission line than would be possible if an ordinary linear dipole were used.

The folded unipole is nothing but a half of a folded dipole, with ground replacing the other half. It is represented in fig. 95, which shows a simple single fold unipole antenna. To analyze this antenna and its operation, let us assume a generator voltage, E_1, and proceed to determine the current, I, flowing in the lower end of part d_l, of the antenna. Roberts[49] has shown an analysis of this type of antenna. Figure 96 is the equivalent circuit of fig. 95, for purposes of the analysis. Also assume that the antenna is 90 electrical degrees tall.

By use of three generators shown in fig. 96, we can by the principle of superposition add the currents produced by each generator to find the current in the lower end of part d_1. The polarities of the generators are shown by the arrows. Since each generator is producing the voltage E_1, and generators 2 and 3 are opposed, the lower end of part d_2 is at ground potential. Generators 1 and 3 produce a voltage $2E_1$ which is impressed on the lower end of part d_1. In view of these facts fig. 96 is in fact equivalent to fig. 95.

As the next step in the analysis, assume for a moment that generator 3 is shut down, and as a result there is only a voltage $2E_1$ developed by generators 1 and 2 between the lower ends of parts d_1 and d_2. It was stated previously that the antenna is 90° tall. Thus, inasmuch as parts d_1 and d_2 form a transmission line of 90° length which is shorted at its upper end, the input resistance between d and d_2 for this transmission line case is very high, and only a very small transmission line current will flow through generators 1 and 2, and into the lower end of part d_1.

Next, assume for a moment that only generator 3 is producing any voltage, and that generators 1 and 2 are shut down. In this case, the lower ends of parts d and d_2 are effectively connected together at the same potential, as far as generator 3 is concerned, assuming zero internal impedance in the generators. Thus parts d_l and d_2 act merely as a 90° vertical radiator (in parallel). Assume that the radiation resistance of this vertical radiator is R, and that it is being driven by generator 3, which is supplying a current equal to E_1 / R to this parallel combination of conductors d_1 and d_2. But since there are two conductors the current, $I = E_l / R$, divides between them. If the conductors d_1 and d_2 are of the same diameter, the current will divide equally between them. Thus I_l, the current is part d is equal to $1/2 \times E_1/R$, and I_2 the current in part d_2 is the same.

Now if all three generators, 1, 2 and 3, are working at the same time, the voltage impressed at the bottom end of part d_l is $2E_1$ and the current in it is $\frac{1}{2} \times E_l/R$ plus a very small amount of transmission line current produced by generators 1 and 2 in series as described above. Since Ohm's Law applies, and resistance equals voltage divided by current, the input resistance at the lower end of part d_1 is:

$$R = \frac{2E}{\frac{1}{2} \times \frac{E}{R}}$$

which is equal to approximately $4R$. This is true provided that parts d_1 and d_2 are close together. The impedance transformation (for equal size conductors d_1 and d_2) is approximately 4.

This transformation ratio can be expressed in the form of an equation:

$$\rho = \frac{Z_1}{Z_o} = (1+\eta)^2$$

where Z_1 = input impedance of the folded unipole antenna.

Z_o = input impedance of a single element depole.

η = input current ratio $I_2\ I_1$

(For equal size conductors this equals 1.)

[49]Roberts, W. V., "Input Impedance of a Folded Dipole." *RCA Review*, Vol. 8, No. 2, June 1947, p.289.

Unequal Conductors

So far in this analysis the case of equal size conductors has been discussed. Now, however, with the transformation ratio expressed in terms of the current ratio (which depends on conductor size), we can consider the case of unequal size conductors, which is shown in fig. 97. As in the previous case, generators 2 and 3 are alike and produce equal and opposing voltages E_2, which place the bottom end of part d_2 at ground potential. However, generator 1 must now be producing a voltage E_1 of such value that no current will flow through generator 3 when generator 3 is not providing any voltage. In this case E_1 is not equal to E_2, and its determination is one essential part of the problem. The division of the antenna current between parts d_1 and d_2 (when generators 1 and 2 are not generating and generator 3 is generating) is the other part of the problem. Roberts uses an electrostatic or capacitive method of analysis[49], whereas Guertler uses another more complicated method.[50] The former described here. Simply stated, the currents will divide directly in proportion to the ratio of the capacities to ground of the two elements, whereas the voltage ratio will be the inverse of this. Thus, in the analysis we can assign capacities C_1 and C_2 to parts d_1 and d_2. Referring to fig. 97 then:

$$\frac{E_2}{E_1} = \frac{C_1}{C_2} \quad or \quad E_2 = E_1\left(\frac{C_1}{C_2}\right)$$

Considering the whole thing as a vertical antenna of parallel conductors, the current I_1 at the bottom end of part d_1 will be:

$$I_1 = I \times \frac{C_1}{C_1 + C_2}$$

where I is the total antenna current due to generator 3.

Inasmuch as parts d_1 and d_2 are a 90° shorted section, the transmission line current, which is very small, can be neglected for practical purposes, and the total current due to generator 3, which we called I, is equal to E_2/R, where R is the radiation resistance of the two parts connected in parallel. The input resistance at the lower end of part d_1 (the driving point resistance of the folded unipole) is:

[50]Guetler, R. "Impedance Transformation in Folded Dipoles," *Proceedings of IRE,* Sept. 1950, p. 1042.

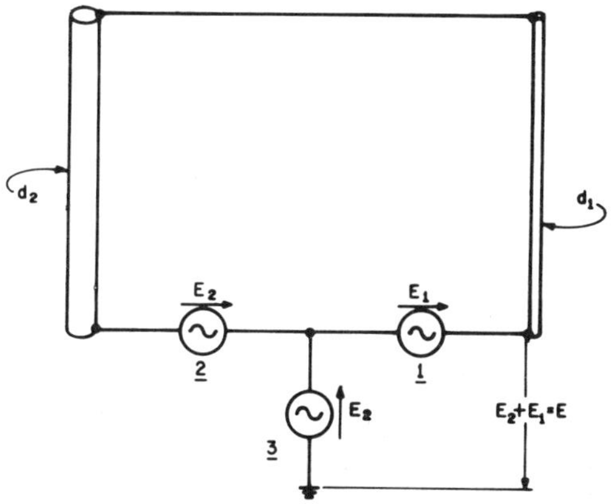

Fig. 97 — Equivalent circuit, for analysis, with unequal diameter elements.

$$R_1 = \frac{E_2 + E_1}{I_1}$$

From this we can derive:

$$R_1 = R\left(1 + \frac{C_2}{C_1}\right)^2 \quad or \quad \rho = \left(1 + \frac{C_2}{C_1}\right)^2$$

This last equation says that the resistance (or impedance) step-up ratio is proportional to the ratio of the conductors' diameters. (Their capacities are proportional to their diameters.) The step-up ratio is inversely proportional to the diameter of the driven fold or element, and directly proportional to the diameter of the grounded element or tower. The spacing between fold and tower enters into the picture slightly also, but it is not critical.

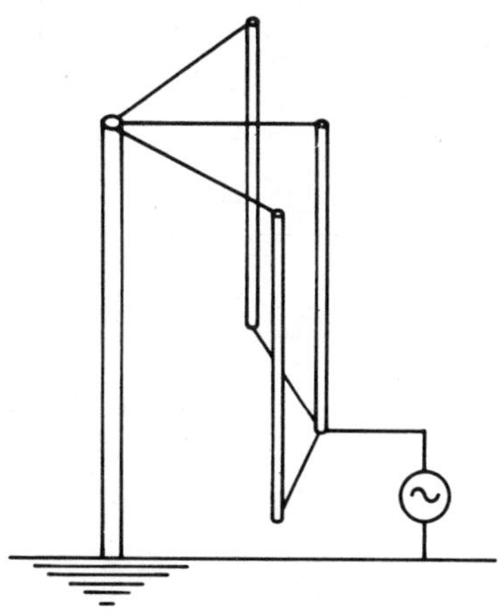

Fig. 98 — Multiple fold configuration for increased bandwith.

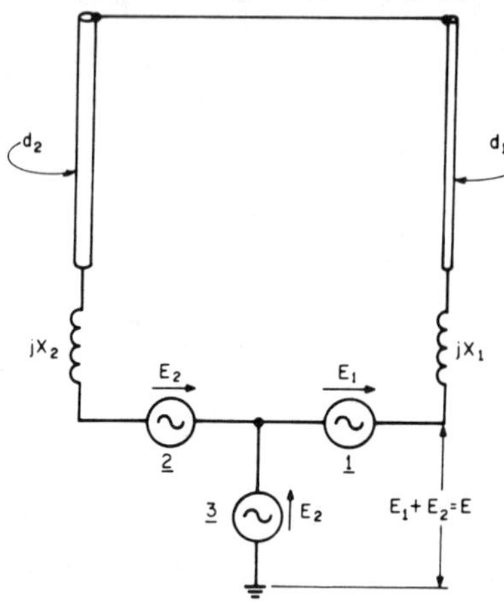

Fig. 99 — Equivalent circuit, for analysis, of the short folded unipole.

This antenna has fairly good bandwidth, but the best way to increase the bandwidth is by increasing the number of folds, as in fig. 98.

It can also be shown that the transformation ratio is:

$$\rho = \left(1 + \frac{Z_1}{Z_2}\right)$$

where Z_1 is the characteristic impedance of a two-conductor transmission line made up of conductors of the smaller diameter, and Z_2 is the characteristic impedance of a two-conductor transmission line made up of conductors of the larger diameter, with the spacing of each of these two lines equal to the center-to-center distance of the two conductors of the antenna.

It is assumed in the above equation that the tower is grounded, and the smaller conductor (or fold) is fed. The impedance step-up will always be greater than four in this case.

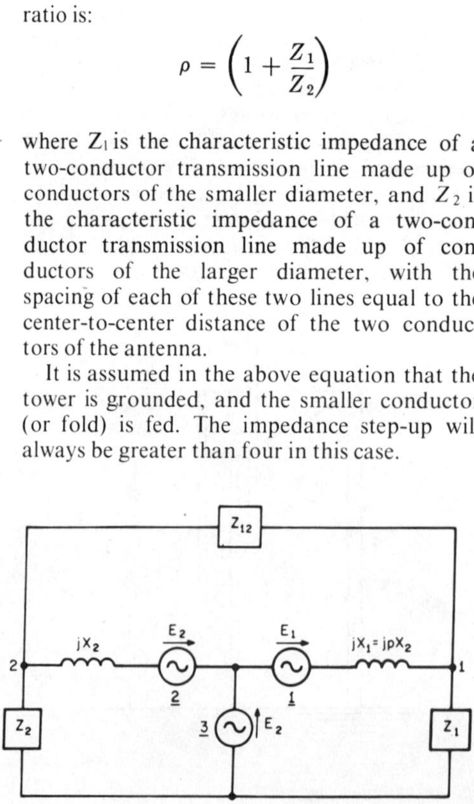

Fig. 100 — Equivalent schematic of fig. 99.

Short Unipole

Thus far a folded unipole 90° tall has been discussed. Now we must turn to the case of the short folded unipole, something much less than 90°, which is the case where the method of feed gives us the greatest advantage of impedance step-up, enabling us to get rid of the helix with its inherent losses. Some assume that the transformation in the short case is the same as that in the 90° case but this is not correct, because in the short case the transmission line currents present become of appreciable magnitude and can no longer be neglected.[51] In this case, fig. 99 applies. Reactances have been arbitrarily added to the two branches of the circuit of fig. 97 for the sake of the analysis. As in the previous case, we can consider the whole thing as an antenna consisting of parallel conductors fed by generator 3, and also we must consider the transmission line case where generators 1 and 2 are in series feeding the short section of transmission line. The circuit of fig. 99 can be replaced by an equivalent circuit as shown in fig. 100. The sections of the unipole have been replaced by boxes containing certain impedances. Z_1 and Z_2 are the impedances to ground of the two parts of the unipole, and Z_{12} is a complex impedance representing a mutual impedance between the two parts (or elements).

As the first step in this analysis, consider the antenna mode, where generators 1 and 2 are not producing voltage, and the configuration is being excited by generator 3. The current, I, will divide into two parts, I_1 and I_2, in accordance with the transformation ratio as previously explained, for $Z_1/Z_2 = p$. If the currents I_2 and I_1 are to be in phase and of the ratio $p = I_2/I_1$, which they must be in the antenna case, then jX_1 must equal $p \times jM_2$. As far as Z_{12} is concerned, no current flows in it, since points 1 and 2 (lower ends of parts d_1 and d_2) are at the same potential and parts d_1 and d_2 are effectively in parallel.

Now to consider the transmission line mode, we assume generator 3 is not producing. In order for the currents in Z and Z_2 to be equal and opposite, which they must be in this case, E_1 must equal p times E_2. Since E_1 plus E_2 equals E (refer back to fig. 95). E_2 is equal to $E/(1+p)$, and E is equal to $Ep/(1+p)$. This is shown in fig. 101.

We are now ready to examine the input impedance of the unipole, at feed point F, fig. 101. The total current flowing at F is the sum

[51] Leonhard, J, *et. al.*, "Folded Unipole Antennas," *IRE Transactions on Antennas and Propagation.* Vol. AP-3, No. 3, July 1955, p. 111.

of the antenna mode current and the transmission line mode current. Let the former be designated by I_A and the latter by I_T. By shorting out the lower generator 3 we have the sum of generators 1 and 2 driving the unipole as a section of transmission line. I_T is therefore determined from the following equation:

$$I_T = \frac{\left[E\left(\frac{\rho}{1+\rho}\right) + \frac{E}{1+\rho}\right]}{j\,(X_2 + \rho X_2 + X_L)} =$$

$$\frac{E}{j\,[X_2\,(1+\rho) + X_L]}$$

where X_L is the transmission line inductance of the shorted section.

The antenna current is found by shorting out the two upper generators 1 and 2. Since points 1 and 2 are then at the same potential, the two conductors of the unipole are considered to be combined into a single vertical radiator, whose input impedance we shall designate by $R_a - jX_a$. The reactance will be negative because the radiator is shorter than 90°. The values of R_a and X_a can be determined by consideration of the radiator's L/D ratio and its electrical length, and by consulting fig. 18A and 18B. We also have X_a and pX_a in parallel in this antenna mode, in series with $R_a - jX_a$. The parallel combination of X_2 and pX_2 equals $[p/(1+p)]X_2$, and the driving voltage for the antenna mode is $E/(1+p)$.

The total antenna mode current is, therefore:

$$I_{A\,(TOTAL)} = \frac{\left(\frac{E}{1+\rho}\right)}{\left[R_a - jX_a + j\left(\frac{\rho}{1+\rho}\right)X_2\right]}$$

Only a fraction of this, however, flows in part d_1 of the unipole. That fraction is:

$$I_A = \left(\frac{1}{1+\rho}\right) I_{A\ TOT}$$

or:

$$I_A = \frac{E}{[R_a(1+\rho)^2 - jX_a(1+\rho)^2 + j\rho(1+\rho)X_2]}$$

From feedpoint F of the unipole to ground, the input impedance is:

$$Z = \frac{E}{I_A + I_T}$$

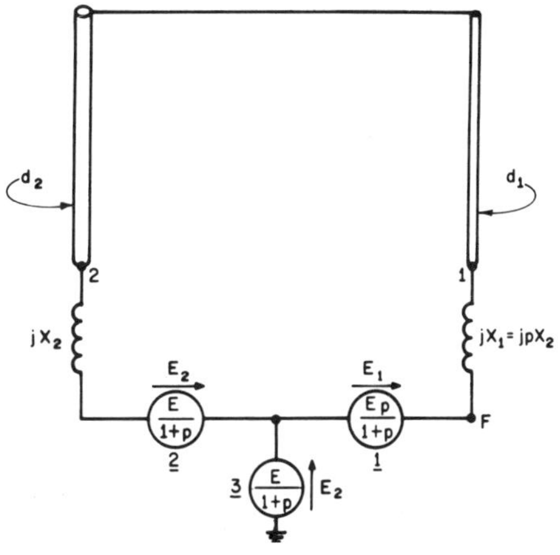

Fig. 101 — Equivalent circuit of the folded unipole showing the generators in terms of **E** and **P**.

From this let us now subtract the value of jX_1 (equals $j\,p\,X_2$) to obtain the true input impedance:

$$Z_{IN} = \frac{E}{I_A + I_T} - j\rho X_2$$

By substituting the values of I_A and I_T in this equation, we can find the input impedance as a function of X as shown in the equation at the bottom of the next page.

The equivalent circuit of the short folded unipole is shown in fig. 102. In the case of the 90° folded unipole, the circuit of fig. 102 and the preceding equation can be used, and since in that case X_a is zero (the antenna is self-resonant), X_2 is zero, and X_1 is infinite (shorted quarter wave line), the resultant $Z_{IN} = R\,(1+p)^2$, which would be expected. Thus the general case does fit the 90° special case.

From all of this we observe that the unipole multiplies the input resistance and reactance of

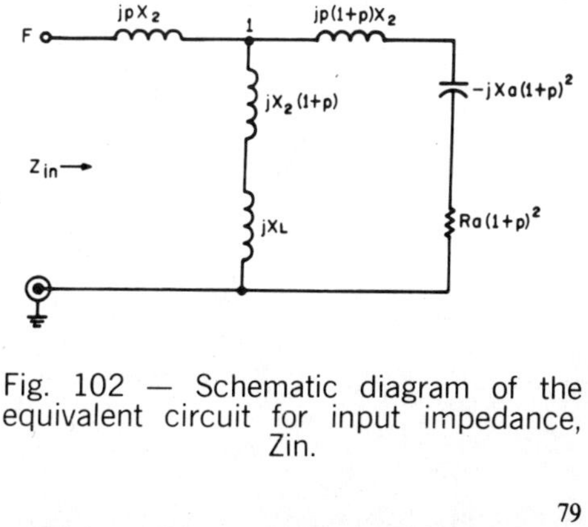

Fig. 102 — Schematic diagram of the equivalent circuit for input impedance, Zin.

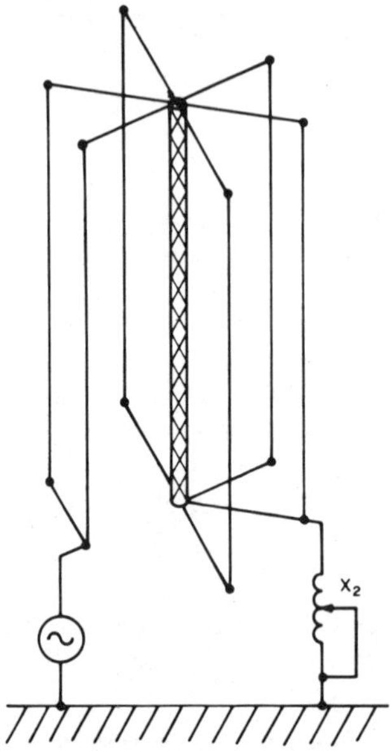

Fig. 103 — "Fattened" or broadbanded tower fed as a folded unipole with multiple fods.

the antenna by the factor $(1 + p)^2$, and it also transforms these new values of R_a and X_a through an equivalent circuit T-network action due to its inherent transmission line mode. The final result is dependent on the value assigned to X_2. In the case of the grounded base short tower, which is quite common, X_2 is of course absent from the circuit. However, use of X_2 can be quite beneficial if optimum efficiency and bandwidth are desired.

Summary

To summarize, the input impedance of the short folded unipole is influenced by anumber of factors. First, it depends on the division of currents between the fed and unfed conductors, on the presence of considerable transmission line currents in the fed and unfed conductors, and on the value of an impedance which can be connected between the unfed conductor (or tower) and ground. The reactance of a short vertical antenna can be reduced by increasing the diameter/length ratio of the antenna. This is done by dropping wires down from outrigger arms connected to the top of the tower, and connecting them in parallel with the tower. This is in effect a "fattening" or broadbanding action, as described in Chapter Eight. Such a structure can be easily modified to the folded unipole method of feed by driving one or more of the drop wires, while the rest are either left connected to the tower or grounded through a reactance X_2. Such a configuration is shown in fig. 108.

An additional advantage of the folded unipole is that the antenna structure is at d.c. ground potential, and no static discharge devices are required. If X_2 is not employed the tower is actually grounded, and lighting circuits, transmission lines to other antennas such as u.h.f. and v.h.f. arrays, and rotator control circuits can be run right up the tower without any isolation. ■

$$Z_{in} = \frac{1}{\dfrac{1}{[R_a(1+\rho)^2 - jX_a(1+\rho)^2 + j\rho(1+\rho)X_2]} + \dfrac{1}{j[X_2(1+\rho)+X_L]}} - j\rho X_2$$

12

THE EFFECTS OF EARTH ON THE EFFICIENCY OF RADIATION AND THE VERTICAL PATTERN

The effects of earth on the efficiency of radiation and the vertical patterns to be expected from a vertical antenna are often misunderstood or are not understood at all. The author discusses these effects in two phases. The first deals with the earth near the antenna and the need for a good ground system. The second is the effect of the earth in the reflection zone on the shape of the vertical pattern.

THE author's antenna articles in *CQ* have brought much mail, most of which has been from readers interested in using vertical antennas on 75 or 160 meters. Power increases authorized on 160 meters, plus the increased availability of s.s.b. equipment for that band, have increased its "population." Because of space restrictions, most of the stations on that band are using verticals of one form or another. There has also been some mail from amateurs on 10, 15, 20 and 40 who are interested in using verticals meters for DX contacts. One of these amateurs announced that he had made the discovery that he could tune and operate a half wave vertical without a ground system, driving it either by a parallel tuned tank or an "L" network whose lower end is grounded. He claimed that since a thermocouple ammeter in the ground lead showed no current, he could dispense with the ground system and its losses! He stated that although I have covered the current fed antennas very well, I should now "discover the new world of the half wave vertical with no ground system." Actually, he was voicing a popular misconception about ground systems and the need for them. This moved me to write this Chapter to present a rather complicated subject in as clear and simple a fashion as possible.

There are two areas of interest when one

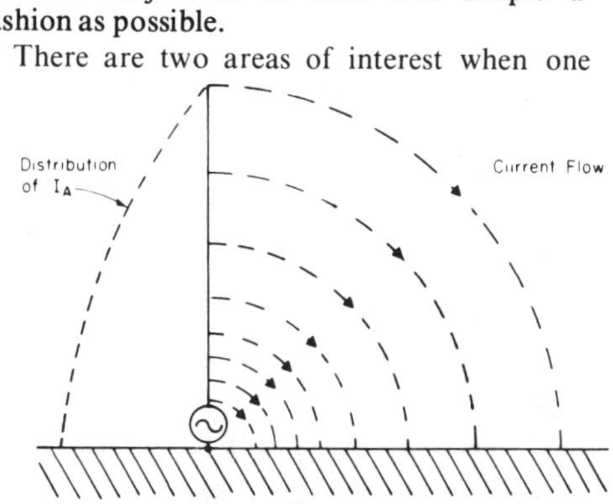

Fig. 104 — Current distribution along a quarter wave vertical and flow of r.f. currents into the ground.

Frequency (kHz)	Depth (meters)		
	$\sigma = 5\times10^{-11}$ $\varepsilon = 81$	$\sigma = 1\times10^{-13}$ $\varepsilon = 10$	$\sigma = 1\times10^{-14}$ $\varepsilon = 5$
10	2	50	150
100	0.67	15	50
3000	0.20	5	17
10000		2	9

Table IV — Depth of penetration for various frequencies and various ground conductivities. ε is the dielectric constant and σ is the ground conductivity in electromagnetic units (e.m.u.).

studies the effect of the earth on radiation from vertical antennas. One of these is the area immediately beneath and surrounding the antenna, and the losses which occur therein which make a ground system necessary. The other area is that of the reflection zone, or as it is called by those of us in the business, "the Fresnel Zone." I shall discuss the close-in area and ground systems first.

Earth Conductivity

There have been many studies made and many papers written on the subject of earth currents and ground systems and their effects on antenna radiation efficiency. G. H. Brown, whom we have met before in this series, performed an excellent analysis and confirmed it by practical experiments.[52] [53] R. C. Hills, G3HRH, a well known radio engineer in his own country, wrote a very fine paper on the subject covered by this Chapter. The matter is also covered quite thoroughly[54] in technical documents of the International Radio Consultative Committee (C.C.I.R.) of the International Telecommunications Union (I.T.U.), Geneva, Switzerland.[55]

All materials are conductors of electricity. Some are very much better than others, and some are very much poorer than others. The latter are sometimes called insulators. The earth is a conductor, and one may find different levels of ability to conduct for the various geological types of earth surface. When current flows in the earth's surface, the well-known "skin effect" occurs.

Radio engineers who deal in h.f. and v.h.f. work have to take this "skin effect" into consideration in the design of conductors and inductors. Simply stated, the current tends to concentrate near the conductor's surface, with the depth of penetration being less at higher frequencies. Usually, conductors are of a homogeneous nature. However, the earth is not homogeneous, being made up of various geological layers. It has been found, for example, that v.l.f. and e.l.f. waves propagate very well through certain types of geological formations, while they are rapidly attenuated by others. At frequencies of 2 mc or more of the current will flow in the upper layer of soil of good conductivity, rather than through underlying strata.

Depth of penetration is a function of the conductivity, and it is greater for poorer conductivities. Table IV shows depth of penetration.[55]

It may be seen that depth of penetration is also inversely dependent on frequency.

The permittivity[56] of the earth as a dielectric also has an effect. Below about 2 mHz it is not important, but at the higher frequencies it becomes increasingly important in considering the earth's role as a reflector. Generally, higher permittivity is associated with higher conductivity, and therefore the effects of earth is usually spoken of as depending on its conductivity.

In the first area of interest, that near the vertical antenna, the earth acts as a return path for the flow of r.f. currents. Let us consider first the case of the quarter wave vertical antenna, current fed at its base. Figure 104 shows

[52] Brown, G. H., "The Phase and Magnitude of Earth Currents Near Radio Transmitting Antennas," *Proc. of IRE,* Feb. 1935, p. 168.

[53] Brown, G. H. et al, "Ground Systems as a Factor in Antenna Efficiency," *Proc. of IRE,* June 1937, p. 753.

[54] Hills, R. C. "The Ground Beneath Us," *RSGB Bulletin,* June 1966, p. 375.

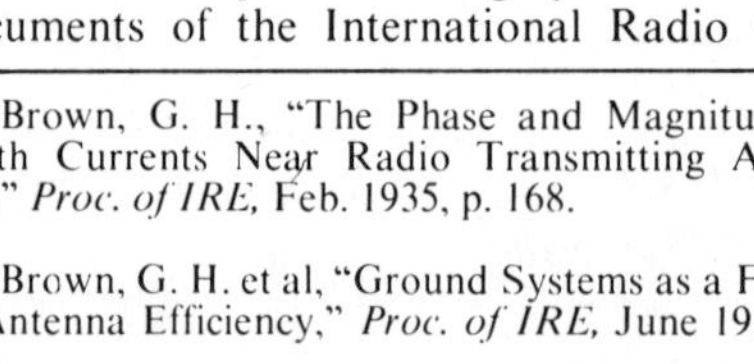

Fig. 105 — Equivalent circuit of fig. 104. The radiation resistance is RA, conductor losses RL, and the effective ground resistance is RE.

[55] Determination of the Electrical Characteristics of the Surface of the Earth," Documents of the IXth Plenary Assembly, Los Angeles, 1959, Vol. 111-Reports, Report No. 139, p. 267.

[56] Permittivity is the property possessed by a material to permit an electric force field to be set up in it with greater or lesser effectiveness. If a material has a high permittivity an electric field will produce more effect in it than it would in a material of low permittivity.

the current distribution along the antenna, and the flow of r.f. currents into the ground. The current in the antenna induces charges in the earth surrounding it, which give rise to the circulating current which flows back to the generator. This flow of current is at a depth of penetration which, as mentioned above, depends on frequency and on ground conductivity decreasing as both increase. The effect of this can be represented schematically by the series circuit of fig. 105, which shows r.f. power being dissipated in the several resistive elements of the circuit, which are the radiation resistance, R_A, the ohmic conductor losses, R_L, and the effective earth or ground resistance, R_E.

The standard practice is to make the radiation resistance as high as possible while making losses as low as possible, thus giving the highest efficiency. Ohmic losses can be kept low by proper design of inductances and use of low loss capacitors. Earth resistance, on the other hand, is something which is inherent to one's location, and unless one wishes to move the home and family to a new location, chosen on the basis of earth resistance, one has to live with it as it happens to be and make the best of it.

There is something one can do to make the best of it and that is to reduce the earth losses by the use of a ground system. Brown determined the distribution of earth currents and earth losses both analytically and experimentally.[52] [53] The distribution around a quarter wave vertical antenna is shown in fig. 106. It may be seen that the current and the losses are highest in the region of greatest current density which is close to the base of the antenna, as would be expected.

Half Wave Antennas

The current distribution on a half wave antenna, and the flow of currents into the ground from it, are shown in fig. 107. The series circuit schematic of fig. 105 applies in this case also. Here again it is necessary to keep the earth losses and ohmic losses as low as possible. The distribution of earth currents and losses around a half wave vertical antenna whose base is close to earth is shown in fig. 108. Brown determined experimentally[53] that the region of maximum current and loss occurs at a distance of about 0.35 wavelengths from the base of the antenna. There is zero loss at the base of the antenna itself, inasmuch as there is no base current because the antenna is fed at a current node. A thermocouple ammeter in the ground lead at this point will read *zero*. However, don't be deceived by this phenomenon, because a thermocouple ammeter in the antenna lead will also read zero.

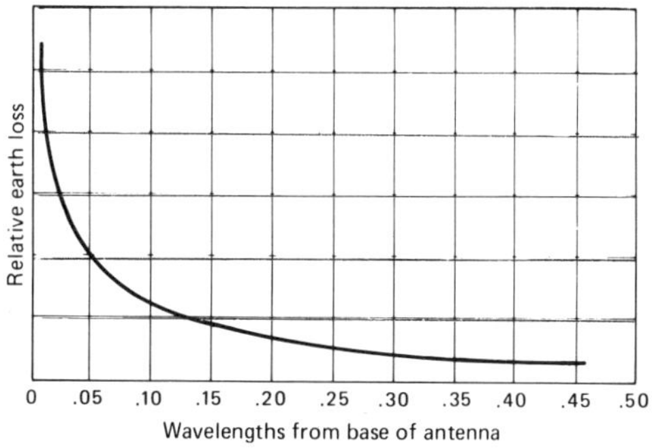

Fig. 106 — Distribution of earth losses plotted against distance, in wavelengths, from the antenna base of a quarter wave antenna.

Efficiency Versus Earth

Figure 109 shows the variations in antenna radiation efficiency over earth, for several values of effective earth resistance and various antenna heights (lengths of vertical radiator). It may be seen from this figure that with low effective earth resistance provided by a good ground system, the short vertical radiator (one eighth wave or so), can be quite efficient. It may also be seen that for a given effective earth resistance the efficiency depends on the antenna's radiation resistance, which for a short antenna is less than that of a tall antenna. There is not too much difference between a half wave and a quarter wave antenna provided that the effective earth resistance is low. There is considerable difference between anten-

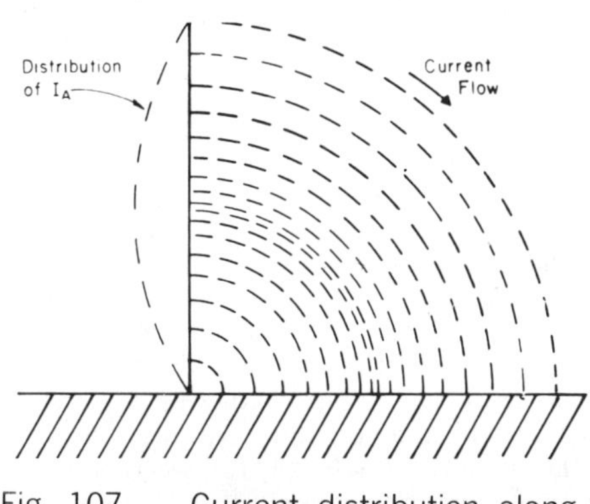

Fig. 107 — Current distribution along a half wave vertical antenna and the flow of r.f. currents to ground of a halfwave antenna.

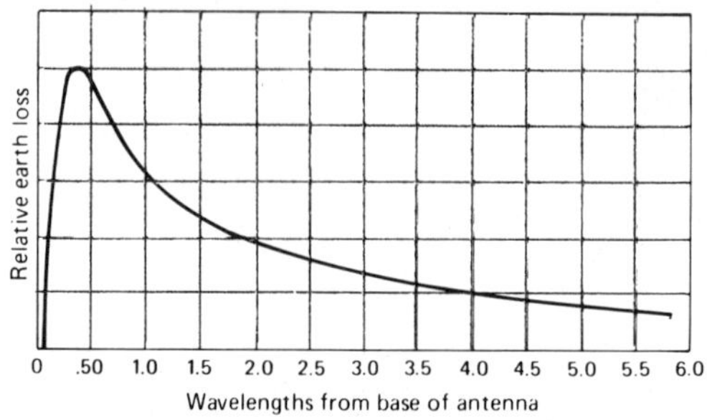

Fig. 108 — Distribution of earth losses plotted against distance, in wavelengths, from the antenna base.

nas of various heights when effective earth resistance is high. This set of curves very effectively shows that low effective earth resistance provided by a good ground system is an *absolute necessity* for vertical antennas of *any* height, if good radiation efficiency is desired.

The correspondent's claim that one does not need a ground system under a half wave vertical radiator is true *only if he is content to throw away from 40 to 80 per cent of his radiated power in the form of earth losses.* He stated, "The ZL's *call me,* when I use my half wave vertical!" This is not surprising, in view of the fact that the half wave's vertical pattern has a lower main lobe angle than a quarter wave would have, and lower than that of the usual horizontal Yagi array. However, he would hit the ZL's even harder if he would put in a ground system. Of course, the half wave vertical antenna is not dependent on a ground plane, however lossy or efficient, for the condition of *resonance,* since it is *resonant in itself* because of its half wave length. However, *it is dependent on a ground plane for its efficiency of radiation,* as is any vertical antenna. Actually, the 5/8 λ vertical is better than the half wave vertical from the standpoints of low angle radiation and feed point impedance.

The Ground Plane

How does the ground plane function? What should be its configuration? To answer those questions, one should look back at figs.104 and 107, which show the necessity for a return path to the base of the antenna, and at fig. 105, which shows the effective series circuit. Since the currents have to flow from all points in the earth surface to a common central point (base of the antenna), it naturally follows that a *radical* configuration of conductors is required. Remembering what was previously said about the depth of penetration of earth currents being inversly proportional to frequency, and recalling that losses in the earth increase with frequency, it follows that the radial wires should be buried close to the surface, for h.f. work. In case of a lawn or other sodded area, let them be just below the level of the sod, at two or three inches depth. One of the half-round lawn edger blades, with a long handle, is ideal for making a slot in sod. By moving the handle laterally as one steps on the blade, the slot may be opened and the sod pushed aside for an inch or so, and the wire dropped or pushed into it with a narrow stick of wood. The wire should be at least number 16 in size. Belden No. 8012, #16 tinned copper, in 1000 foot rolls, makes excellent ground radial wire. This quantity will make 25 radials of average length of 40 feet.

How many radials are required? Reference should be made to fig. 14 and 15 to note the effect of varying the number and length of radials. Refering to fig. 109, a ground system of at least 120 radials a half wave long would be required to give an effective earth resistance of from 2 to 5 ohms. In the case of a short radiator, it is more important to have a large number close in, as shown in fig. 110. If the configuration of one's property limits the length of radials in a particular direction, a larger number of radials should be laid down where they have to be short than where they can be longer, to improve the return path for the earth currents in the "short" sector, to keep the losses down. In the case of the half wave radiator, however, there need be no larger number close-in, because the current density in that region is low. Instead, there should be a large number *as long as possible,* preferably a half wave long, for lowest losses and greatest efficiency, bearing in mind that the region of greatest current density occurs from about 0.35 wavelengths from the antenna on out to further distances.

Since not everyone has enough property for such long radials, it could actually turn out that a quarter wave antenna with a large number of short radials might be more efficient, although its angle of main lobe radiation would be higher. If one were fortunate enough to be located on a body of water, with its inherent high conductivity, one would indeed have a very fine site for an antenna of any height, with a low loss ground plane.

Fresnel Zone

The second area of earth influence is that of the reflection zone, or "Fresnel Zone." Here again the earth conductivity plays a part, although not such a great one as it does in the launching of the wave in the immediate area of the antenna. In practice, all antennas must be installed at some finite distance above ground, or with base on the ground. Therefore their vertical radiation patterns are always influenced by the earth, and never, never conform to the patterns for "free space" conditions. The signal radiated at any angle is the vector sum of the direct ray and the reflected ray. Consider fig. 111, wherein a dipole is located at a distance *h* above ground. For horizontal polarization:

$$E'_\theta = 2\ E_\theta \left(\sin \frac{2\pi\ h}{\lambda}\right) \sin \theta$$

where: *Eθ* is the resultant field at a distant point,
$E\theta$ is the field at the same point, in free space,
θ is the angle of elevation of the point, and
h is the height of the center of the dipole above ground in wavelengths.

$$E'_\theta = 2\ E_\theta \left(\cos \frac{2\pi\ h}{\lambda}\right) \sin \theta$$

These two equations are of the same basic form, the difference between them being that in the case of horizontal polarization the wave is reflected with a phase shift of 180° at the reflection point, and with vertical polarization there is zero phase shift. *This is only with a perfectly conducting reflecting plane,* which is never the case in actual practice, although sea water approaches it.

In the "real life" case, with an imperfectly conducting earth, there is an effect called the "Brewster Effect," named after an English physicist, Sir David Brewster, who investigated certain optical properties of reflecting planes as related to polarization of incident light waves. Using non-polarized light, Brewster found that it was partially reflected and partially refracted at a plane surface, such as the surface between air and glass. (You snorkel fisherman may be aware of this effect as you look upward from below the surface of the water.) He discovered that there is a critical angle at which the reflected wave is totally plane polarized. In the case of radio waves and the earth's surface, it is the horizontal component which is reflected at the critical angle, and the vertical component which is reflected or suppressed. See fig. 112. If the earth were a perfect dielectric, the critical angle would be 15° above the horizontal. However, it is not a perfect dielectric nor is it a perfect reflector, and energy will be lost at the reflection point into the earth. The ground reflection coefficient and the Brewster Angle will therefore vary

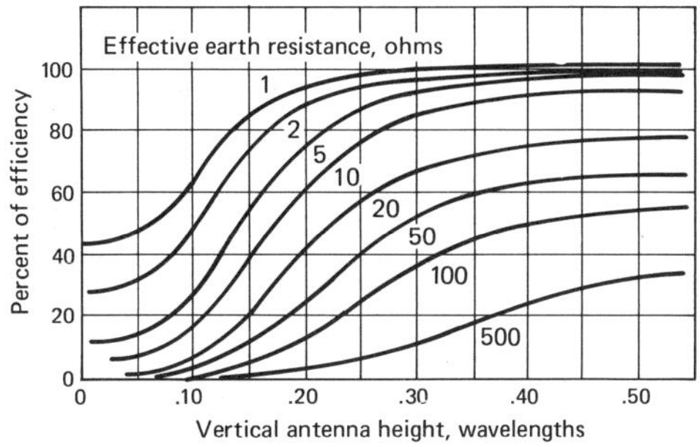

Fig. 109 — Efficiency of radiation for antennas of different heights over various values of earth resistance.

with ground conductivity.

Figure 113 shows the phase and amplitude of the ground reflection coefficient for varying angles of incidence of a vertically polarized wave. The curves for higher frequencies may be taken to represent the trend for poorer conductivity at the lower frequencies. Figure 114 shows the same thing for a horizontally polarized wave. The following relationship applies:

$$\textit{Reflection Coefficient} = \frac{(\textit{Reflected Wave})}{(\textit{Incident Wave})} \underline{/\phi^\circ}$$

What do these curves show us? We can easily see that there is a "suck-in" or attenuation of the low angle radiation from a vertical antenna, plus a large phase shift at the reflection point for very low angles of incidence. On the other hand, with the horizontal antenna there is very low attenuation at the reflection point, but there is almost 180° phase shift at all angles of incidence. This tells us then that the horizontal antenna, to have any good low angle radiation, must be located at considerable height above the earth (at least a half wavelength and preferably higher) to equal the low angle performance of the vertical. This

Fig. 110 — Radial pattern for a vertical antenna.

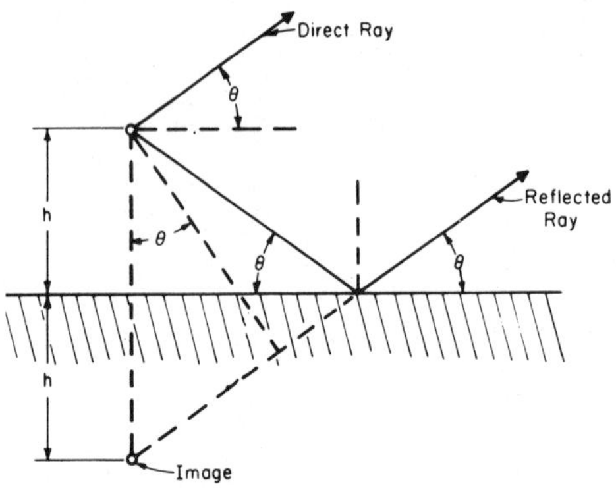

Fig. 111 — Relationship between the direct and reflected wave from a dipole located a finite distance above ground.

fact, coupled with mechanical and structural considerations plus those of available space, make the vertical an ideal choice for the three low bands, 40, 75 and 160 meters. There is another benefit to be gained, also, and that is the fact that with a good ground plane and a quarter wave antenna whose maximum current is at the base of the antenna, the rays that account for most of the radiation from the antenna emanate from its lower portion of high current, and their reflection points will be close-in due to the geometry of the situation. If the ground system is extensive enough (a quarter wave radius), they will, for the most part, fall on the ground system itself, thereby incurring less loss than if they were to fall at some distant points beyond the control of the station owner. In fact, if one were to assume that the center of radiation of the quarter wave vertical is at a very small distance above ground, for the sake of the formulas and

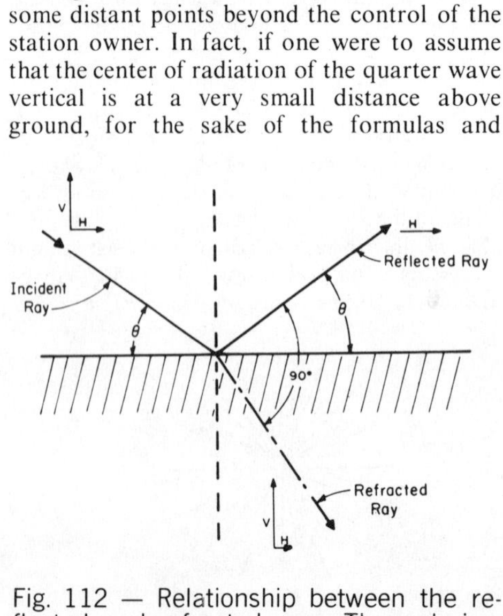

Fig. 112 — Relationship between the reflected and refracted rays. The polarizations are shown for each ray.

curves above, the results would not be very far from reality as far as the vertical pattern shape is concerned. There would be some suck-in at low angles, which would depend to a great extent on the excellence of the ground plane itself. With a computer one can actually compute the vertical plane patterns from antennas of various heights, taking into account the various ground conductivities which exist in different areas of the country.

This might all be summed up by saying that *ground systems are very important,* and that the conductivity of the earth itself is also a vital factor in antenna efficiency. The following points should be observed, when installing a vertical radiator:

a. Install as many radials as possible, as long as possible, in your ground system.

b. A short radiator can be quite efficient if ground radials are of sufficient number and length to keep earth losses low.

c. If you use a halfwave or 5/8 wave radiator, use radials of at least one half wave length, if possible, to keep earth losses low and to enable you to realize the full benefit of the added height and its increased low angle radiated field.[53 55 57]

d. Radials should be at least #16 copper wire, and should be brazed or soldered to a copper strap of 2″ by 1/16″ size around the base of the antenna. If the antenna can be located on top of a sheet of copper or expanded copper screen, about two or three feet square, this is ideal as a tie point for the radials.

e. Soft or hard solder may be used. If it is used, all soldered joints should be coated with an asphalt paint or compound to prevent them from corroding when buried. Asphalt roofing cement or tile cement, when heated to make it more fluid, is ideal for this. A soldering torch should be used for soldering. Do not rely on the limited heat from a soldering iron. Brazing or silver solder is best.

f. Do not cross radials or tie their distant ends together. This avoids circulating currents which could cause loss.

g. If a Gamma Match type of feed is used with a grounded tower, the connection between tuning unit and tower should be heavy because it will have to carry considerable current if the feed point resistance is low.

h. Radials should be buried about 2 or 3 inches, just enough to get them under the sod for protection.

[57]HF Vertical Plane Patterns of Monopoles and Elevated Vertical Dipoles With and Without Extended Ground Systems," *Naval Electronics Laboratory Center Report* 1567, 25 June 1968.

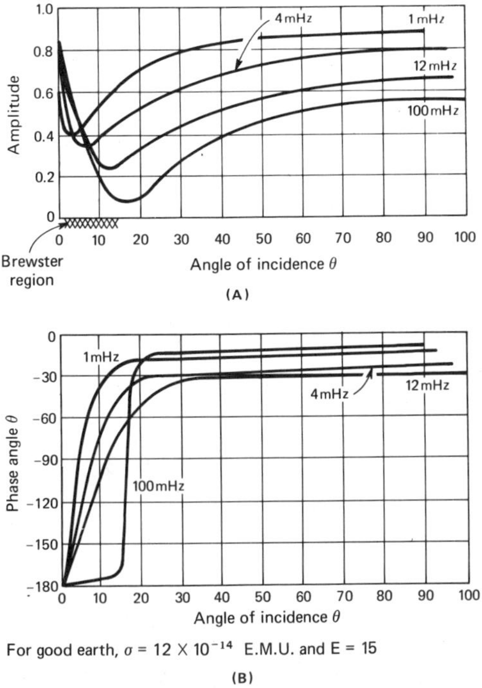

Fig. 113(A) — Amplitude of the ground reflection coefficient for various angles of incidence at different frequencies. (B) Phase angle for various angles of incidence at different frequencies. Both plots are for a vertically polarized wave. For good earth conductivity:

$$\sigma = 12 \times 10^{-14} \text{ e.m.u. and } \varepsilon = 15.$$

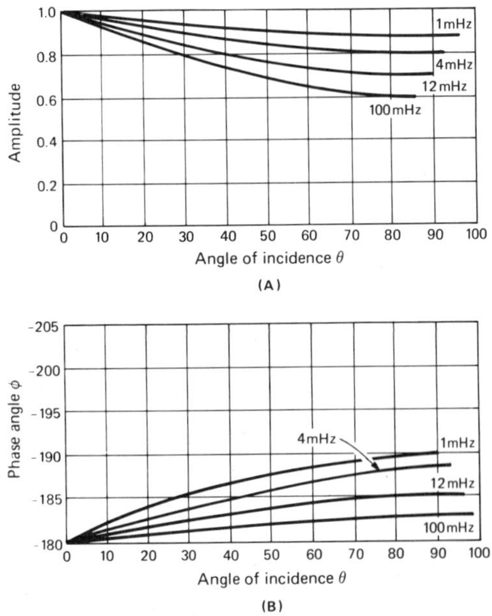

Fig. 114 — (A) Amplitude of the ground reflection coefficient for various angles of incidence at different frequencies for horizontally polarized waves. (B) Phase angle for various angles of incidence at different frequencies for horizontally polarized waves for good earth conductivity.

Summary

Let me relate some personal experience. When I started operation in Maryland in 1959, I installed 8 radials of #9 aluminum wire, and several years later added 8 more. In the intervening years the aluminum gradually corroded and was eaten away, and I was not sure how much ground system I actually had left. Not wishing to dig it up to find out, I bought 2000 feet of #16 copper and proceeded to install 50 new radials, of varying lengths from 35 to 80 feet, at intervals of about 7° around the Mark IV Antenna. I had been working on 160 meters with the old system with fair results, but was not satisfied with signal strengths from the more distant stations. There had seemed to be some loss in the system. Upon connecting in the new radials I noted an immediate improvement in signal strengths, both on receiving and transmitting, especially with the stations in more distant areas. The improvement was of the order of three or four S-points. An electrically short antenna which the Mark IV is (about 0.16 wavelengths tall) on 160 meters, can be quite efficient when earth losses are kept low. ■

13

EVOLUTION AND CONSTRUCTION OF A SIMPLE FOUR-BAND VERTICAL RADIATOR, THE MARK I

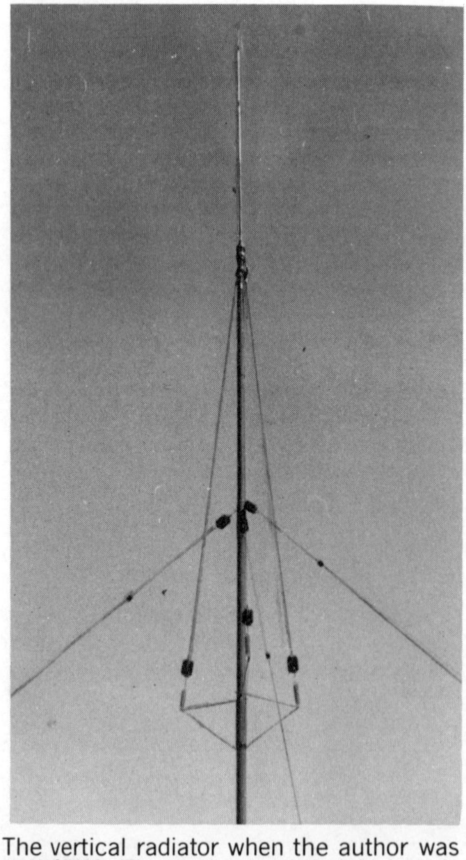

The vertical radiator when the author was at W4RXO. The total over-all height is only thirty-nine feet. Three drooping radials are used on 21 mHz, and a ground screen on all other bands.

IN the preceding chapters of this book I have dealt with the theoretical and design aspects of vertical antennas. In this chapter I am going to describe a simple four-band vertical which was actually built in 1952 when I was W4RXO at Isle of Palms, South Carolina. My location was about 200 yards from the beach, and it provided an excellent take-off for low angle signals across the Atlantic and south to the Caribbean. I previously had several separate antennas for the 14 and 21 mHz bands, and wanted one antenna which could be used on those two bands plus 7 and 3.9 mHz.

The Mark I antenna described here is a combination of two vertical radiators. One of them covers the 21 mHz band, and the other covers the 14, 7 and 3.9 mHz bands. I neglected the 28 mHz band at that time because we were in the part of the sunspot cycle wherein the 28 mHz band would not be very useful for a few years. For 21 mHz I had been getting quite good results with a drooping ground plane, so decided to use that concept for the 21 mHz portion of this antenna. For the 14 mHz band I chose an overall height which would be 0.58 wavelength, which would give a good low angle pattern and a fairly favorable base impedance. The overall height includes the 11 foot length of the drooping ground plane's whip, making the total overall height 39 feet which is 0.58 wavelength at 14 mHz. This length becomes 0.30 wavelength at 7 mHz and 0.16 wavelength at 3.9 mHz.

Radiation Patterns

The calculated pattern for a 0.58 wave vertical radiator is shown in Fig. 115(A). The cal-

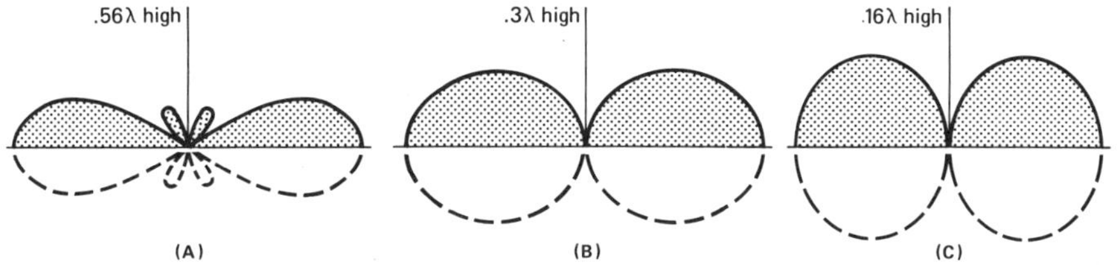

Fig. 115 — Theoretical vertical radiation patterns. The antenna is considered to be 0.58 wavelength on 20 meters, 0.30 wavelength on 40 meters and 0.16 wavelength on 75 meters. Radiation on all of these bands is at the most favorable angles.

culated patterns for 0.30 and 0.16 wavelength vertical radiators are shown in figures 115(B) and 115(C) respectively. An inspection of these patterns will show that this vertical gives good low angle radiation.

Constructional details of the Mark I are shown in fig. 116. The mast is made from 2″ aluminum tubing. It rests upon a large and rugged standoff insulator, and is guyed at a point 20 feet from the ground, in three directions. The guys are broken up by "egg" insulators, so that no portion of any guy is as long as 11 feet.

The fitting which holds the 21 mHz whip at the top of the 2″ mast is a surplus type *MP-48* whip base insulator, complete with spring. It provides a means for fastening the coaxial feedline internally. The whip is made of surplus *MS53, MS52, MS51* and *MS50* whip sections, with the last section cut to give an overall whip length above the base insulator of 11 feet. A machined fitting holds the base of the *MP-48* whip base tightly inside the top of the 2″ mast. The drooping ground plane radiator is fed by means of *RG-8/U* coaxial line which goes down through the mast, and which terminates in an *SO-239* chassis type female connector set into the mast just above the base insulator. The outer conductor of this line is thus connected to the mast at top and bottom.

The three drooping ground plane radials are mechanically fastened to the *MP-48* whip base by means of three small "egg" insulators, and are electrically connected to the *MP-48* base at only one place, which is a bolt in the side of the "ground" portion of the *MP-48* base, above the spring. The length of each radial is exactly 11 feet, measured from this bolt. The bottom ends of the radials are secured to large "egg" insulators, which are fastened to brackets which hold the bottom ends of the radials a distance of 1 foot out from the mast.

At this point you are probably beginning to wonder how the 11-foot whip at the top of the mast is made to act as extension of the mast, and to radiate, at the three lower frequencies. Well, take a look at fig. 116 and you will see that I used a 7-foot 6-inch piece of *RG-8/U* coaxial line as a shorted quarter-wave stub cut for 21 mHz, the open end of which is connected between the whip and the drooping ground plane. The outer braid of the stub line is connected to the mast at several places along its length. The connection between inner conductor, outer braid, and mast itself, at the lower end of the stub, serves to effectively

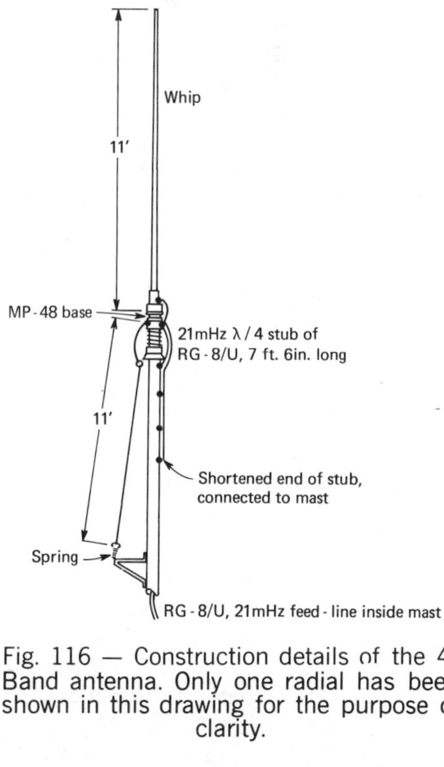

Fig. 116 — Construction details of the 4-Band antenna. Only one radial has been shown in this drawing for the purpose of clarity.

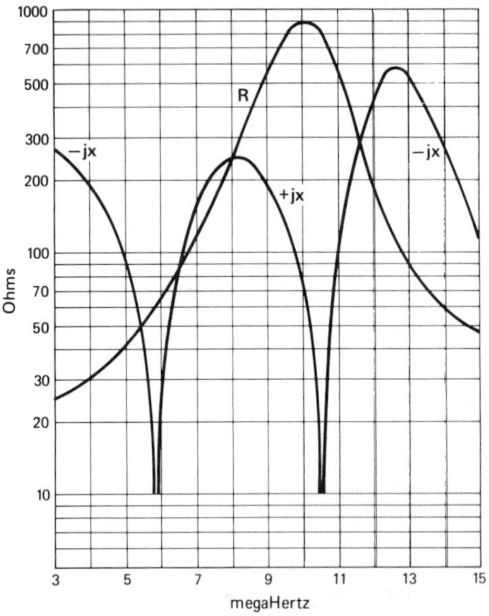

Fig 117 — Measured resistance and reactance of the 4-Band antenna. The method of matching was developed after these measurements has been made.

connect the whip to the mast at the three lower frequencies. This would not be true, of course, if the stub were insulated from the mast.

The Matching Network

Now, the question of feeding power to this antenna at four different frequencies arose. Obviously, I needed some sort of matching system between the base of the mast and the *RG-8/U* transmission line for the three lower frequencies, and I also had to feed 21 mHz power to the drooping ground plane radiator at the top of the mast.

C_1 = 350 mmfd variable (E. F. Johnson type 350 D 70).
L_1 = 13 turns 3/8" OD copper tubing, 4" diameter, 7" long.
S_1 = 2 section, 4 position ceramic switch.

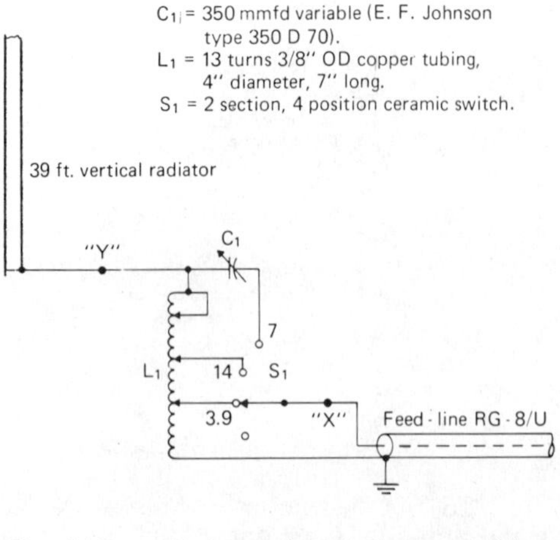

Fig. 118 — Schematic of the matching network.

The first step was to measure the radiation resistance and reactance of the antenna at 3.9, 7, and 14 mHz with a *General Radio Model 916A* r-f bridge. In order to get the complete picture, I took measurements at one megaHertz intervals from 3 to 15 mHz, and the results are plotted in fig. 117.

The base impedance of the antenna at the three lower frequencies is shown in this table:

	R	*X*
3.9 mHz	30	-j200
7.25 mHz	135	+j190
14.25 mHz	59	-j195

These three impedances can be matched to the 50-ohm line by a network such as the one shown in fig. 118. At 3.9 mHz the large inductance is used as a tapped loading coil. The exact size of the coil can be calculated, using an inductive reactance of 200 ohms at 3.9 mHz which is required to balance out the 200-ohm capacitive reactance of the antenna. The position of the tap can be determined easily by experiment. At 14 mHz the tap on the coil is at a higher position. At 7 mHz we have a simple *L*-network, with the series capacity in the feedline to balance out the inductive reactance of the antenna at this frequency. Adjustment of the 14 mHz tap, and the setting of the variable capacitor for 7 mHz may be easily done by trial.

How did I feed the 21 mHz power to the coaxial line up the mast? No, I didn't just plug it in when we wanted to go on 21 mHz. I *wound the large coil out of coaxial line,* and fed the 21 mHz power through this line, and used coaxial line from the coil to the base of the mast as the lead-in for the three lower frequencies! Take a look at the sketch, fig. 119, and at the photographs, and you will see how this was done. Where did I get the copper coaxial line? Why, I made it! It was very easy. I took 15 feet of *RG-8/U,* stripped the vinyl covering and outer copper braid from it, thus leaving the polyethylene dielectric and the inner conductor. I then slipped this into a 15-foot length of ⅜" O.D. copper tubing, where it fits very nicely, making the required length of copper coaxial line, and I wound it into an inductance of the proper size. A *PL-259* coaxial male plug was soldered on each end of the coil, and through the use of *PL-258* female junction fittings, connection was made to the feed-line at each end of the coil, with *PL-259* plugs. A similar short piece of this home-made copper tubing coaxial line was used between coil and mast fitting, as the lead-in. I passed it through an *E. F. Johnson Type 135-67* insulator before soldering the *PL-259* plugs on the ends. It fits very nicely through the hole in the insulator, after the bolt is removed. After installation, the hole was caulked with caulking compound to make it water proof.

I used one coaxial line for the 21 mHz and lower frequency feeds from the transmitter. The switching from 3.9 to 7 to 14 to 21 mHz was done by means of a large rotary switch, as shown in fig. 119. Relays may be installed to do this job by remote control, if desired.

Note that the inner conductor of the coaxial coil *L1* is grounded at 3.9, 7, and 14 mHz, at the ground end of *L1,* through the contacts of *SW1.*

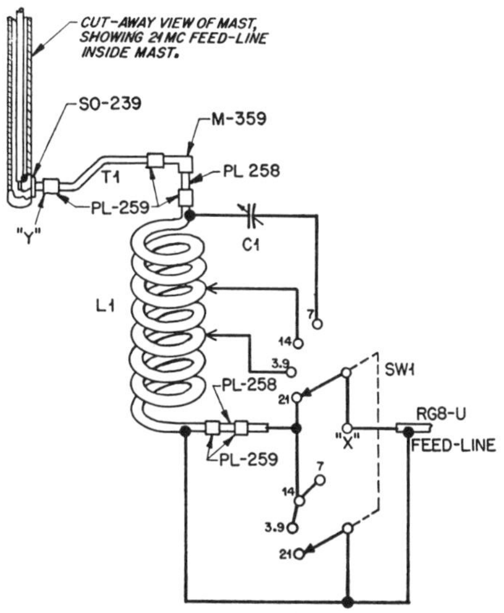

Fig. 119 — Construction details of the matching network.

Ground System

The tuning unit, built into a waterproof aluminum box, is shown in the photographs. Once adjusted, it is closed up and left alone. The rotary switch shaft projects through the side of the box with a rubber grommet to keep water out. It is turned by using a screwdriver.

An antenna of this kind needs a ground system. I buried eight radials, of #12 bare copper wire, each 60 feet long. They were all tied together at the base of the antenna, and *one* ground lead was brought up to the tuning unit cabinet. A six-foot square piece of copper ground screen would be a very good addition to this ground system. The radials should be soldered to it, as shown in fig. 120. Of course, the ground system was installed *before* the resistance and reactance measurements were made!

In tuning up, it was useful to insert a 0-5 ampere line-current r-f ammeter at point "X," and a 0-9 ampere antenna current r-f ammeter at point "Y." The antenna current meter needed an adapter to enable it to be plugged in between the coaxial lead-in and the connector at the base of the mast. This is shown in fig. 121. For r-f current measurement purposes, the inner and outer conductors of the coaxial lead-in were connected together, as shown. When the system is properly tuned up on any one of the lower frequency bands, the power in the feedline and antenna can be easily computed by Ohm's Law, for the line impedance is 50 ohms, and the resistance of the antenna is known. These two power readings should be equal, and should also be equal to the final amplifier's power input multiplied by a proper efficiency figure. In my case the power *output* was approximately 670 watts.

Results

The results obtained with this antenna system on all four bands were most gratifying. The drooping ground plane gave very good

Base and tuning unit at W4RXO. The two-inch diameter aluminum tubing rests on a surplus insulator.

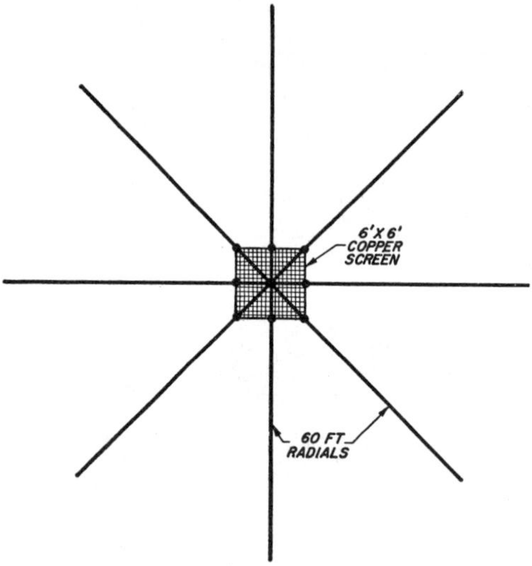

Fig. 120 — The efficiency of this type of antenna depends upon the ground system. This is the suggested arrangement for radials at the base of the antenna.

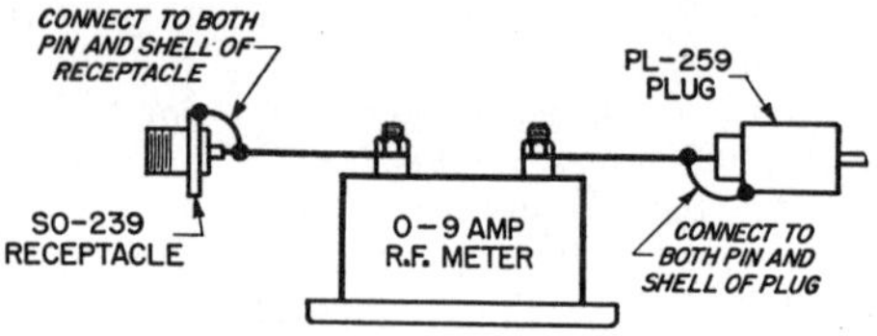

Fig. 121 — This antenna current meter should be inserted at point "Y" in fig. 119 while the antenna is being tuned up.

low-angle radiation on 21 mHz. You may wonder why I "drooped" the radials so much, when the usual practice is to run them out at about 45 degrees from vertical. Well, I wanted to lower the angle of radiation, for one thing. And then, too, I wanted to keep them down alongside the mast so that they would not have a tendency to act as top-loading at the lower frequency bands and thus throw off the antenna height pattern calculations. By installing them as shown, there is a slight mismatch between the *RG-8/U* line and the ground plane antenna, but the standing-wave ratio on the line at 21 mHz was only in the order of 1.35:1.0, and I did not consider this to be objectionable. I could perhaps have eliminated standing waves entirely by making adjustments to the shorted quarter-wave stub, but it was not worth climbing a 2″ pip mast to do it!

On 14 mHz I noticed the difference between this antenna and the usual half-wave doublet or quarter-wave vertical, both of which I had used previously. Overseas signals were much stronger, and there were less "short-skip," or "stateside" QRM, because of the shape of the vertical radiation pattern, shown in fig. 115. On 7 mHz it performed beautifully, giving *S9* plus reports on almost every contact. On 3.9 mHz, because it is somewhat shorter than a quarter-wave, its efficiency was reduced, but it put out a very good signal both locally and at distances of several hundred miles. At night on 3.9 mHz I worked stations over 1,000 miles away with this antenna, in the Western U.S., and the West Indies, from W4RXO.

Now for a word about TVI. This antenna definitely helps that situation. In the first place, it is fed by coaxial line, which is buried underground from the transmitter out to the tuning unit. Use of coaxial line enables insertion of a low-pass filter in the line leaving the transmitter. If any harmonics do get by the filter, they are discriminated against by the antenna itself, because as we go above 0.58 wavelengths in length, the radiation from a vertical antenna shifts from low to high-angle, where it can't do any harm to your neighbor's TV picture. The 21 mHz drooping ground plane also discriminates against harmonics because of the mismatch which occurs at harmonic frequencies between antenna and feed-line.

And there you have it — the Mark I four-band antenna which takes very little space, which is neat in appearance, and which is easily erected. Comments and questions from readers will be welcomed. ■

14

THE FOUR BAND VERTICAL DX ANTENNA, MARK II

WHILE on duty at Charleston, South Carolina in 1952, I designed the Mark I antenna described in Chapter Thirteen. The article on that antenna brought a flood of mail from amateurs all over the world, which proved to my satisfaction that the vertical antenna is a subject of great interest. When I moved from Charleston, I left the antenna with a local amateur, vowing that someday I would build another one and build it better. A move from an apartment location to a new house with a large yard in 1960 gave me the opportunity and the Mark II version, described herein, is the result. The reader will quickly note the utter simplicity of construction of the Mark II, and will also be charmed by its low cost. Its appearance is very attractive also, which is a bit of an inducement for approval of the XYL, and it arouses only mild interest from the neighbors.

Review

A bit of review of the theory of vertical antennas is in order. A vertical antenna 0.625 λ (⅝ λ) in height is at optimum for maximum low angle radiation and minimum high angle skywave. This well known fact is borrowed from standard broadcast antenna design practice. Inasmuch as my main interest was then in the 14 mHz band, I designed the Mark II to be 0.625 λ high for that band, which makes it 41 feet in height. On 7 mHz it is 0.31 λ high, and on 3.9 mHz it is 0.174 λ high. The two latter values are reasonable ones for those bands, and provide adequate coverage. Calculated vertical radiation patterns for various antenna heights are shown in fig. 122.

I wanted to make the Mark II self-supporting, which meant that I could not use guys as ground plane radials for a 21 mHz antenna on

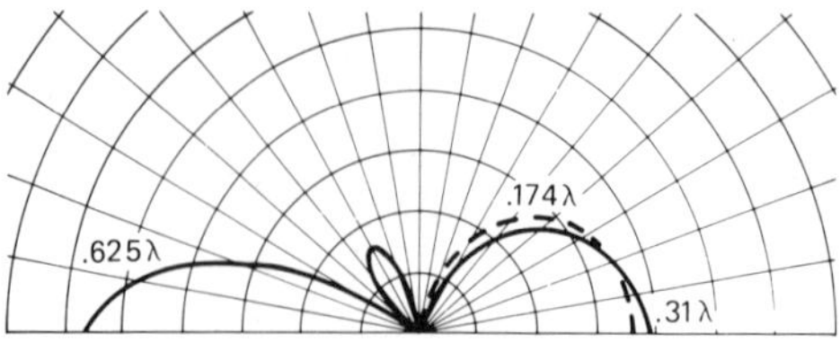

Fig. 122 — Calculated vertical radiation patterns for various antenna heights

top as in the original version. I was not completely satisfied with the drooping ground plane anyway, as it left a lot to be desired in ease of matching and adjustment. Also, the surplus MP-48 base insulator for the 21 mHz antenna made the whole thing rather top-heavy. The final design of the Mark II is very clean and attractive, as may be seen from fig. 123. The lower 20 feet of the mast is a piece of *Sears and Roebuck* aluminum irrigation pipe, Catalog No. 42HRM5978, found in their Farm Equipment Catalog. This pipe has a wall thickness of 0.04 inches. The upper 21 feet is made of three sections of *Reynolds* "Do It Yourself" tubing obtainable from most hardware stores. I used three sizes, 1¼ inch, 1 inch, and ¾ inch. These sizes do not telescope together with a tight slip fit. I accomplished the tight slip fit by cutting 6 inches off one end of the ¾ inch and 1 inch sizes, splitting each 6 inch piece down one side with a hacksaw, and spreading it open slightly so that it fits tightly

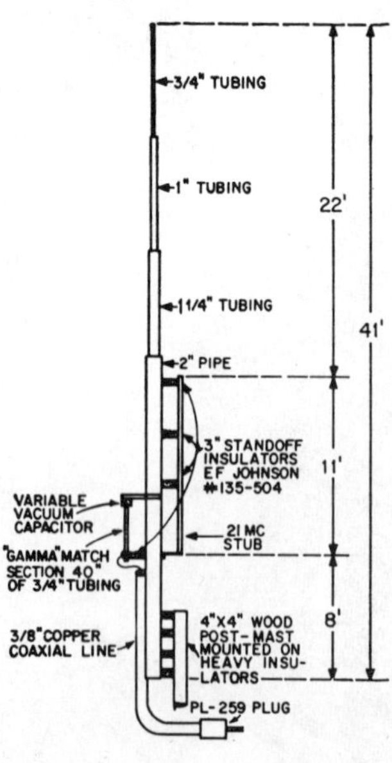

Fig. 123 — Final design of the Mark II vertical.

around its parent section as a bushing 6 inches long. This then takes up the slack and enables tight slip fits to be made between the three upper sections. Brass bolts, 8-32, are used to fasten these two splices. The upper 21 foot portion is then fitted into the lower 20 foot length of 2 inch pipe by means of four spacer bars made from 5/16 inch aluminum stock, one foot long, ¾ inches wide, and filed to fit the inner curve of the 2 inch pipe. Figure 124 shows the details of this splice. Four ¼-20 brass bolts 2½ inches long are used to fasten this splice.

The whole antenna weighs approximately 11 pounds, and when held horizontally by one person at the bottom end, the 41 foot length sags only about 2 feet. Yet it is quite flexible, and it will bend with a high wind and will not break. It is supported on four heavy standoff insulators made by *Lapp Insulator Co.* They are side-mounted on a wooden 4 × 4 post which is set 3 feet into the ground, and extends 5 feet above ground. The stand-off insulators are spaced 15 inches apart. The details of a mounting, which is done with a galvanized U-bolt muffler clamp and a short length of 1 inch aluminum angle, are shown in fig. 125.

21 mHz Stub

The 21 mHz stub on the side of the mast is made of an 11 foot 4 inch length of ¾ inch tubing, with the bottom 4 inches flattened and bent to form the mounting bracket, which is secured to the side of the mast by four self-tapping screws. Three 3 inch *E. F. Johnson* standoff insulators are used to hold the stub rigidly to the mast. Each insulator is first mounted on a strap of 16 gauge aluminum which goes around the 2 inch mast, and then the stub is secured to the other end of the insulator by means of a screw through the ¾ inch tubing. Three insulators are required to hold the stub to the mast. This prevents changes in spacing as the mast sways in the wind, which would result in detuning on 21 mHz. This stub which is connected to the mast at a point 33 feet below the top, and the mast itself, make a "J" antenna of the basic type shown in fig. 126. The upper 22 feet of the mast acts as a ½ λ ver-

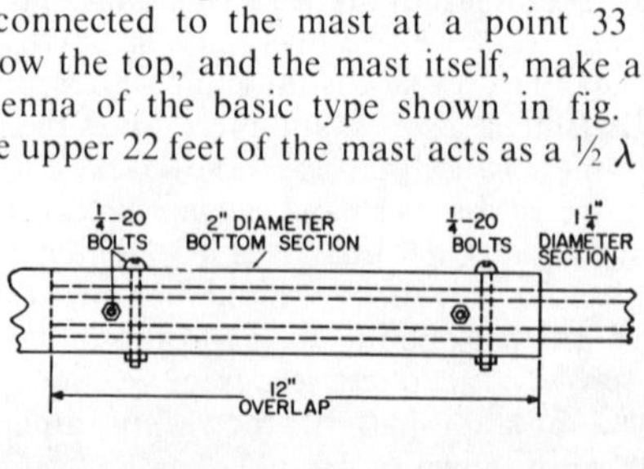

Fig. 124 — Details of the 1¼" to 2" splice. Spacers described in the text, solidify the splice.

Fig. 125 — Photo of the U-bolt mounting. The 1" angle iron is secured to the Lapp #13981 standoff insulator. A 1" × ½" × ¼" spacer bar is placed between the mast and angle iron. (The Lapp Insulator Co. is located in LeRoy, New York.)

tical antenna on 21 mHz, and this is fed by a ¼ λ section, 11 feet long, shorted and grounded at the bottom. The coaxial feedline is connected to the antenna side of the ¼ λ section through a sliding 40 inch "Gamma match" section made of ¾ inch tubing and using a *Jennings* Type UCSL 4-250 mmfd 3 KV variable vacuum capacitor. The mounting of this capacitor is shown in fig. 127. One side of the capacitor is supported by a wrap-around strap bracket made of flattened ¾ inch tubing. The rotor side is fastened to the "Gamma match" tubing. The bottom of the "Gamma match" slides in a fitting mounted on another 3 inch insulator. The coaxial feedline runs down the mast from the bottom of the ¼ λ section, and the outer conductor is connected to the mast every two feet. This copper coaxial line is the feedline for the 21 mHz "J" antenna, and also acts as the lead in for the whole vertical antenna on 14, 7, and 3.9 mHz.

Tuning Unit Construction

I used the same feeding and matching arrangement on the Mark II as used on the original version in Chapter Thirteen. The coaxial line is made of ⅜ inch copper tubing with RG-8/U inner conductor and dielectric pushed through it. This copper coaxial line with polyethelyne dielectric is then wound up in the form of an inductance for the tuning network for the three lower frequency bands. It also acts as coaxial feedline for the 21 mHz "J". The schematic of the tuning network, with circuit constants, is shown in fig. 128. A heavy duty 4 position rotary switch with ceramic insulation is used to shift the incoming feedline to the tap points for 14, 7, and 3.9 mHz, and to the feedline connection for 21 mHz. The flexible tap leads are made from RG-8/U copper braid, and are connected to the inductance by Mueller No. 45C copper clips. The jaws of the clips must be spread slightly to fit the ⅜ inch tubing. The tuning unit is mounted inside a 12 × 22 × 12 inch aluminum box, with a homemade watertight cover which has a rubber gasket cemented around the inside edge. A view of the tuning unit housing is shown in the photograph. The coaxial inductance and the *E. F. Johnson* variable capacitor are mounted on 3 inch standoff insulators. Copper strap ½ inch wide is used for connections. The lead in insulator is an *E. F. Johnson* Type 135-67-2, whose hole is just the right size for a tight fit for the coaxial line. Caulking compound seals the tubing in the hole and seals the insulator to the box. Type PL-259 plugs, with a PL-258 adapter, are used to connect the outgoing coaxial line to the coaxial feedline on the mast. The incoming coaxial line which is buried from the house is connected to the tuning unit through a PL-259 plug which mates with an SO-239 chassis type fitting in the bottom of the tuning unit box. The tuning

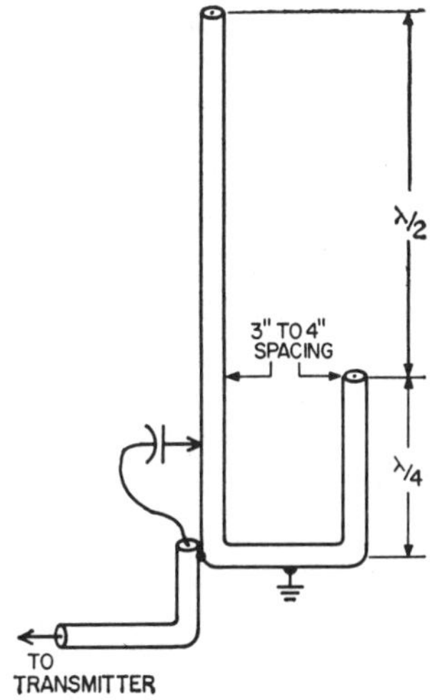

Fig. 126 — Typical "J" antenna. This is formed by the 21 mHz stub. The capacitor shown is the vacuum variable in the Gamma match.

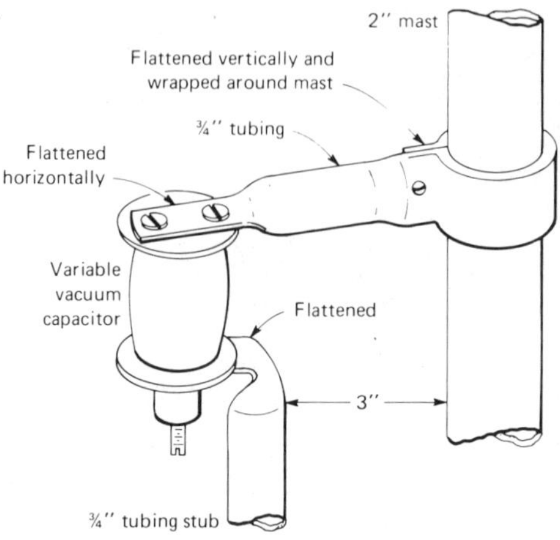

Fig. 127 — Detail view of the Gamma match vacuum variable mounting.

unit is side-mounted on the 4 × 4 post, about a foot above the ground as seen on page 93.

Radial System

There is a buried ground system consisting of 300 feet of #9 aluminum wire *(Sears and Roebuck* clothesline), around the base of this antenna. In our yard I was able to get in two 60 foot radials, plus six 30 foot radials, spaced every 45 degrees. These are all tied together by means of solderless connectors underground at the base of the mast, and two of the radials are left long enough to be brought up to the bottom of the tuning unit where they are connected to a ¼-20 brass bolt. The head of this bolt, inside the box, is connected to the bottom end of the coaxial coil outer conductor by means of a heavy copper strap, which is soldered as well as bolted in place under it.

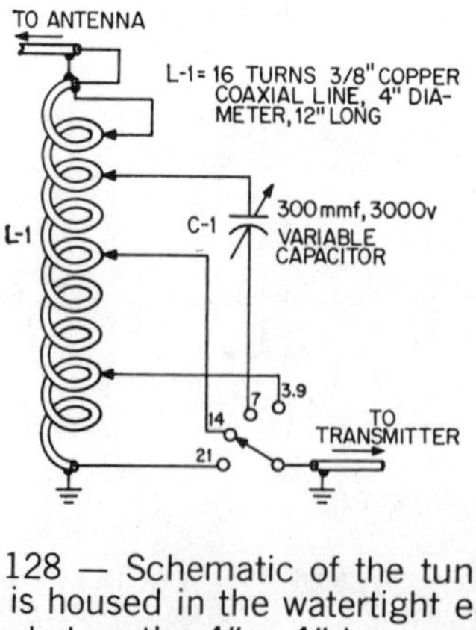

Fig. 128 — Schematic of the tuning unit. This is housed in the watertight enclosure mounted on the 4" × 4" base post shown in the photo on page 93.

Adjustment

Adjustment of this antenna on all bands was done with the aid of a *General Radio* Model 916A Radio Frequency Bridge, using the station *Collins* 51-J2 receiver and LM Signal Generator. A complete frequency *vs.* impedance measurement was made on the antenna alone, and the results are plotted in fig. 129. These resistance and reactance curves conform quite closely to Schelkunoff's predictions in his book, "Antenna Theory and Practice," (pages 438 and 439) wherein he discusses and plots impedances of cylindrical antennas for various values of Ka,

$$\text{where Ka} = 120 \left(\log e \frac{2l}{a} - 1 \right)$$

l is the length of the antenna and a is the radius of the antenna. This is the same formula given in Chapter Three. An average figure is used for the a of this antenna for the computation.

The bridge was connected to the input of the tuning network by means of a PL-259 plug. With the switch in the 21 mHz position, the vacuum capacitor mounting bracket was slid up and down the mast, thus adjusting the length of the "Gamma match", and the vacuum capacitor was tuned, to give an input impedance of about 50 + j0 ohms. This simple sentence sums up about two hours of climbing a ladder, adjusting, climbing down, reading the bridge, climbing the ladder, adjusting, climbing down, reading the bridge, and so on! This is not a simple operation. It requires patience! The final adjustment of the vacuum capacitor is very critical. All I can say is that if you try it, the adjustment probably will be a bit different from mine, but the vacuum capacitor bracket should be about 3 feet up from the bottom of the stub. Start with more length than this, and work down to it slowly. The variable vacuum capacitor is just hung there on its mounting bracket, right out in the weather. After all, it is made of heavy glass, so why cover it? The weather can't do it any harm.

The next operation, after you have recovered from the ladder climbing detail, is to throw the switch to the 3.9 mHz position and adjust the taps on the coil for 50 + j0 ohms input. This inductive loading arrangement is quite frequency sensitive, so pick the frequency you use most. In my case I set it for the middle of the phone band. The feedline tap and the turns-shorting tap are adjusted for 3.9 mHz. The v.s.w.r. rises a bit when the transmitter is tuned to 3.8 or 4.0 mHz, but not excessively (less than 2.5 to 1). The turns-shorting tap is then left alone in all subsequent adjustments.

In my case, I had 14½ turns left in the coil, with the feedline tap 3¼ turns up from the bottom.

It was impossible to obtain exactly 50 + j0 ohms, but I ended up with 48 + j8.6, which was quite close. The 7 mHz adjustment should be done next, with the switch in that position. The tap to the variable capacitor ended up 13¼ turns from the bottom, with the capacitor set at about 40% capacity. This adjustment is quite easy to make, and we arrived at exactly 50 + j0 ohms at 7.25 mHz quickly. The 14 mHz adjustment, intentionally left until last as it is the most critical, was next made, with the switch in the proper position. This tap ended up at 11½ turns from the bottom, with an input impedance of 41-j3.5 ohms, which is quite close. Here again it was impossible to get an exact 50 + j0 ohm match. The v.s w.r. measured at the transmitter end of the 60 foot buried RG-8/U feedline is in all cases less than 1.5 to 1, which is quite acceptable.

The adjustments should be rechecked, beginning with that for 3.9 mHz, to make sure that placement of leads has not changed anything during the tap changing process. It will probably be necessary to touch up each adjustment slightly several times, depending on how critical one is of the results. Also, check again with the cover of the box in place! Distributed capacity changes the settings slightly.

If one does not have an r.f. bridge to do this job, the best suggestion is for use of a grid dip meter and v.s.w.r. meter, with a low powered transmitter to excite the antenna on all bands. An r.f. ammeter can be inserted in series with the lead to the antenna to show antenna current, if desired. The tap adjustments stated above can be used as a beginning, and fine adjustments can be made by the cut-and-try method to get proper loading and lowest possible v.s.w.r. on the input line.

If one wishes, relays may be installed to do the band switching remotely, from the operating position. The control lines should run in direct-burial type of cable, in order to avoid r.f. pickup.

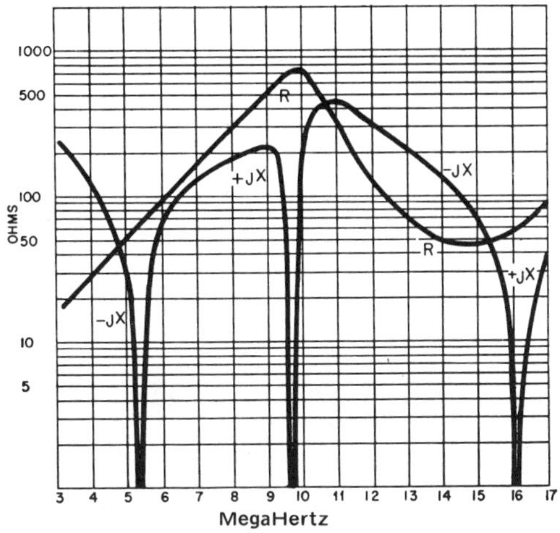

Fig. 129 — Impedance curves for the Mark II.

Conclusions

This antenna performs beautifully for DX work, especially on 21 and 14 mHz, where the radiation is at quite low angles. I noticed a great difference between it and the temporary ½ horizontal dipole used on DX contacts. It reduced stateside interference and brought in the Pacific islands and other faraway places with real sock. Also, it helped reduce TVI because any v.h.f. harmonics that may sneak out of the rig after all the filtering are shot up in the air at high radiation angles, instead of being radiated out along the ground where they can get into a neighbor's TV Set. The Mark II is about as attractive in appearance as an antenna can be, and will hardly cause a raised eyebrow from the XYL especially if you put a 3 foot white picket fence around it to keep the children from touching it when it is "hot." If you let her plant roses around the picket fence, she'll be positively delighted. ■

15

REMOTE CONTROL FOR THE MARK II FOUR BAND VERTICAL DX ANTENNA

One of the drawbacks in having a tuning unit out at the antenna is that long trek out behind the house to switch bands. Here's an idea that will save lots of walking and keep you as snug as a bug in a rug during a long winter storm.

View of the tuner housing with the cover removed showing placement of relays and wiring. It can be seen that the relays are mounted on standoff insulators to insure adequate insulation from the metal housing. The buried cable enters through a hole in the lower right corner of the box.

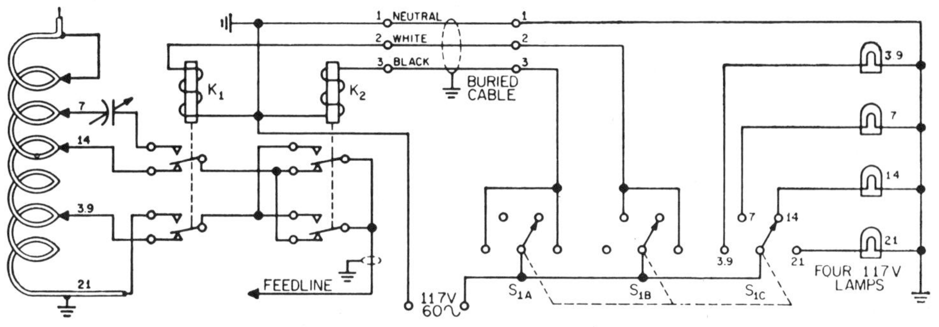

Fig. 130 — Schematic of the remote control unit. The switch and relays are discussed in the text. When both relays are de-energized the tuner is set on 14 mHz. When K only is energized it is switched to 7 mHz. If only K_2 is energized the tuner is switched to 3.9 mHz. When K_1 and K_2 are energized simultaneously the tuner is switched to 21 mHz.

HERE is an added refinement which can easily be incorporated into the Mark II Four Band DX Antenna described in Chapter Fourteen. By means of two fairly heavy duty r.f. relays, an appropriate length of three-wire cable and a small control panel, band switching of the tuning unit can be done remotely from the operating position. This is a very handy modification, as it makes it unnecessary to go out in the snow or rain to throw the band-switch, thus overcoming the only objection to this antenna. I made the modification during the antenna's annual summer overhaul.

Kilowatt Relays

The circuit of the tuning unit, with the relays and remote control system, is shown in fig. 131. The r.f. circuitry was not changed at all from that shown in the previous chapter. However, the original four-position rotary switch was removed, and two relays were installed in the bottom of the tuning unit as shown in the photograph. These relays are double-pole double-throw, Aemco Type 51. They are rated by the manufacturer to handle a kilowatt, and have 115 volt a.c. coils.

Relay Modification

I found it necessary to make a slight modification of the relays to prevent arc-over at the contacts. This modification, shown in fig. 131, consists of raising the normally closed upper contacts by inserting 6-32 nuts under them, providing additional clearance between upper and lower contacts and providing increased armature swing. This is necessary to prevent arc-over at K_2, where high r.f. voltage exists between upper and lower contacts, especially on 3.9 mHz. If you intend to run low power, this modification may not be necessary, but in my case, K_2 did arc at an average kilowatt input to my 4-1000A. Relay K_1 did not arc over, but I modified it anyway, just in case. There should be definite contact pressure in both the normally open and normally closed positions of these relays, with a slight wiping action occurring between fixed contacts and armature, keeping the contacts clean. The spring tension should be increased as much as possible for positive relay action.

It is necessary to mount these relays on small bushings about a half inch high, to provide clearance between the relay and the housing. The screw heads on the terminals of these relays are counter sunk less than 1/16 inch into the bottom of the ceramic base. If the relays are mounted right on the metal housing, arcs and short circuits will result. The phenolic spacers which come with the relays are not satisfactory for r.f. use.

Underground Cable

The cable I used is a 3 conductor plastic covered type, easily obtainable from an electrical supply store, and is designed for underground burial. It has two #14 insulated conductors and

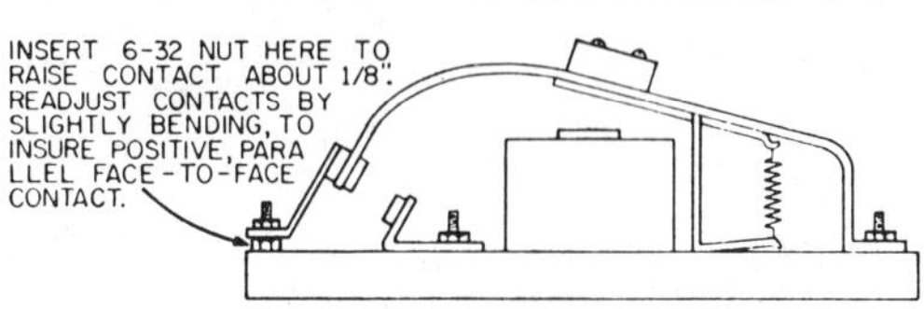

Fig. 131 — Side view of relay showing position of the additional 6-32 nut on the normally closed contact. The additional gap is required for high power installations.

one ground wire. It is buried about six inches underground from the tuning unit to the house. It terminates at the operating position in a standard three-prong polarized twist-lock receptacle which is mounted in an electrical outlet box in the wall behind the desk. One must be sure to connect the ground side of the control circuit to the ground side of the 115 volt power line, or blown fuses will result. In my case, the antenna control circuit was permanently connected to the transmitter control system, and power was applied when the main transmitter circuit breaker was closed.

It will be noted that when both relays are in the de-energized position, shown in fig. 130, the normally closed contacts connect the transmission line to the 14 mHz tap on the tuning coil. This was done intentionally, so that with no power applied to the control circuit I could use the antenna for receiving 20 meters. The 14 mHz band was the one that I used most during daily operations, so it was fortuitously convenient to leave the relays in the unenergized position for that band.

Control Panel

The control panel is a standard 1¾ by 19 inch aluminum rack panel. Four 115 volt indicator lamps are mounted on it, along with a 3 pole 4 position rotary switch. The switch is a miniature type, Grayhill No. 5-003-4. These miniature switches are rated to carry 5 amperes at 115 volts, and are ideal for situations like this one where the 1¾ inch panel does not present sufficient clearance for a larger type. The control panel was mounted in a space, formerly vacant, beneath my single sideband exciter. I wired the switch and lamps so that the lighting of the lamps from left to right corresponds to clockwise switch positions, in order from 21 to 3.9 mHz.

A Word of Warning

It was not necessary to readjust any of the taps on the tuning coil after substitution of the relays for the rotary switch. The v.s.w.r. did not change from its previous value on any frequency. Needless to say, switching should never be attempted with r.f. power applied. Serious arcing and damage to the transmission line and transmitter components could result, especially with high power.

The hole formally occupied by the rotary switch was plugged with a dummy ¼-20 bolt, and the power cable entrance bushing was plugged with sealing compound. The whole tuner housing was repainted and the antenna was again ready for service.

It was a pleasure to be able to switch bands without getting up from the operating console. It certainly was the means of my getting on 7 and 3.9 mHz more often when I happened to tune across those bands and heard old friends talking. Just turn the antenna bandswitch, set the transmitter dials to pre-calibrated positions, and away we go! See you on all bands, fellows!

16

THE MARK III DX ANTENNA

Here is a vertical antenna, 50 feet tall and self supporting, for the 80, 40, 20 and 15 meter bands. Its low angle of radiation makes it ideal for DX operation.

THIS antenna proved itself with such success that it seems worth passing on to those who would like to build a colinear vertical. The antenna was originally intended for use only on 14, 7, and 3.9 mHz, but I found that it can be successfully used on 21 mHz as well. I do not know what its vertical pattern is on 21 mHz, nor do I know what its current distribution is, but it worked.

Antenna Operation

14 mHz Operation

The Mark III antenna is 50 feet tall, and is grounded at its base. It would have been impractical to make this taller and heavier mast self-supporting and still insulated at the base like the Mark II, so I decided to ground it and shunt feed it with several feed rods. Basically it is one half of a three-element center-feed colinear on 14 mHz, as can be seen in fig. 132. (The mirror image below ground is the other half.) A three-element colinear has a gain of 3.2 db over a dipole. Replacing the bottom half with a ground plane and putting all the power into the upper half gives an additional 3 db theoretical gain. The current distribution on this antenna is shown in fig. 132. The phase reversal at the point a quarter-wave above ground can be accomplished in several ways, by a shorted quarter-wave stub, a tuned circuit, or a quarter-wave coaxial sleeve. The latter is shown in fig. 133. This design is used quite commonly at v.h.f. and u.h.f., where size permits several elements with coaxial sleeves to be stacked vertically. The phase reversal occurs at the open or bottom end of the coaxial sleeve. Inasmuch as the current loop in the bottom portion of the mast occurs at ground level, the bottom end can be grounded and fed with a shunt or "gamma" type of feed, as shown in fig. 133. The vertical pattern over ground of average conductivity at 14 mHz is

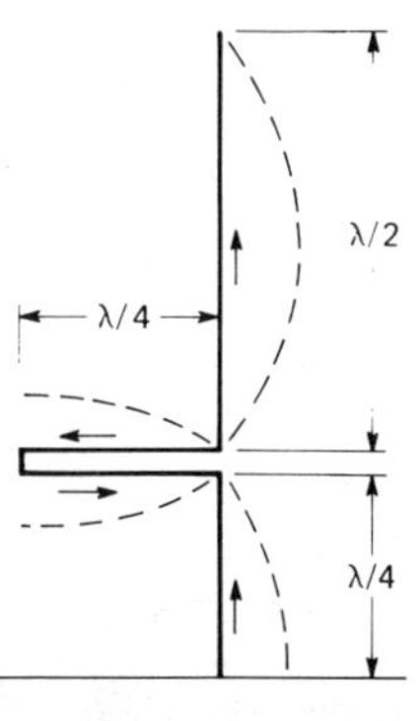

Fig. 132 — On 14 mHz, the antenna operates as one half of a 3 element center fed colinear. Also shown is the current distribution. The mirror image below ground constitutes the second half of the antenna.

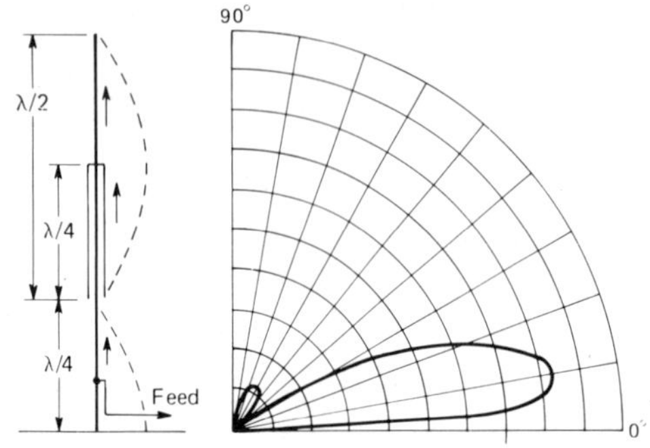

Fig. 133 — Phase reversal is accomplished by the use of a coaxial sleeve rather than the stub shown in fig. 132. The resulting vertical radiation pattern is shown.

also shown in fig. 133. Note that the pattern is "squashed" down, with most of the power going out at angles from about 3 to 22 degrees. This is ideal for DX work because it places the first reflection zone out at a distance of from 1100 to 2100 miles.

7 mHz Operation

On 7 mHz the coaxial stub is too short to resonate, and it has little if any effect on the current distribution or the vertical pattern. The antenna behaves as a ⅜ wave vertical, whose vertical pattern over average earth at 7 mHz is shown in fig. 134. Here again we can feed this antenna with a shunt or "gamma" feed, as shown. Note that this antenna's pattern is useful for angles of about 5 to 45 degrees, which is good for medium and long distance work on 7 mHz.

3.9 mHz Operation

On 3.9 mHz the stub again has no effect,

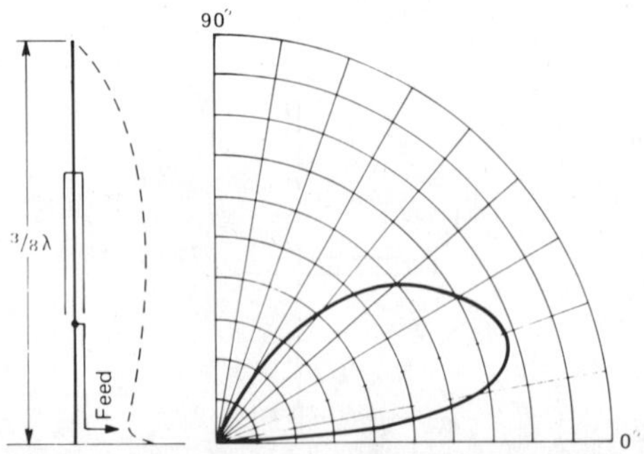

Fig. 134 — Since the coaxial stub will not resonate at 7 mHz the antenna behaves as ⅜ wave vertical and presents the radiation pattern shown above.

and the current distribution is shown in fig. 135. At this frequency the antenna is 0.22 wavelengths tall. The vertical pattern is also shown. Useful power is radiated over vertical angles from 8 to 60 degrees, making this antenna good for short, medium, and long distance work on 3.9 mHz. Here again the mast can be fed against ground with a "gamma" type feed rod. There is only one problem at 3.9 mHz, and that is the "Q" of the antenna. It is slightly too short to be self-resonant (not quite a quarter-wave), and therefor the impedance looking into the bottom of the feed rod is in the order of 9 + j110 ohms. This is with the feed rod connected at the highest possible place on the bottom of the mast, just below the open end of the stub. The "Q" is about 12. With the matching network (to be described later) adjusted to give a v.s.w.r. of 1 at 3.9 mHz, the

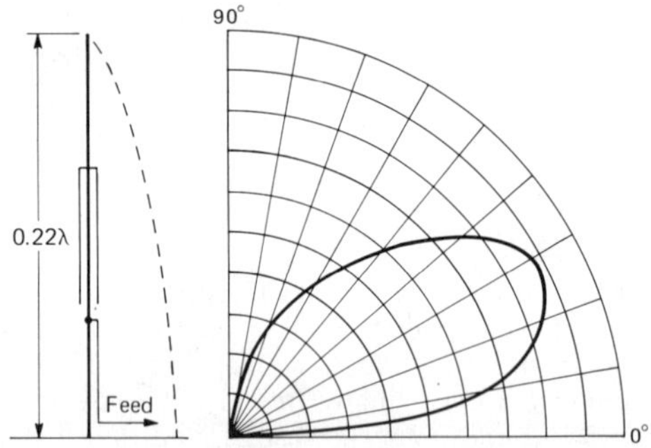

Fig. 135 — The coaxial stub has no effect on 3.9 mHz operation and the antenna acts as 0.22 wavelength unit with the vertical radiation pattern shown.

v.s.w.r. rises to about 1.7 at 3.8 and 4.0 mHz. This is not too bad, however, and can be easily compensated for in the output pi-network of the transmitter. On 7, 14, and 21 mHz, however, the v.s.w.r. is a flat 1:1 over the whole band.

Mast Construction

The dimensions of the mast are shown in fig. 136. The bottom section is a 20 foot length of 4″ aluminum irrigation pipe, of which four feet is set in the ground. The next section is a 20 foot length of 3″ irrigation pipe, with 30″ of this telescoped into the top of the bottom section. The third piece is 10 feet of 2″ irrigation pipe, with 18″ telescoped into the top of the 3″ piece. The top section is a 9 foot length of 1½″ tubing, which is telescoped 12″ into the top of the 2″ piece.

The telescoped junctions are made as shown in fig. 137, by using some hardwood shims which are planed down to the proper shape, wrapped with thin sheet aluminum, and are held in place by four large bolts which pass through the junction (two bolts at right angles to the other two). Bolt sizes are shown in the table in fig. 137.

The coaxial sleeve is hung around the 3″ section. It is 16.5 feet in length, and is shorted to the mast at its top and is open at its bottom. Feed rod dimensions are shown in fig. 138. The 3.9 mHz feed rod, which is connected to the mast at a point just below the open end of the sleeve, is also used for the 7 mHz feed by switching matching networks. The 14 mHz feed rod is connected to the mast about 8 feet from the ground. The connecting point for the 21 mHz feed rod was found by experimentation to be about 4.5 feet from ground. The three feed rods are made of 1″ aluminum tubing, and are connected to the mast by brackets of aluminum strap at their upper ends, and are held by standoff insulators at their bottom ends. Details of these fittings follow in a later paragraph. Switching between feed rods to change bands is done in the tuning unit at the base of the mast. Those who read the chapters on the Mark II antenna will recognize this tuning unit box and its wood post, which has been cut off short.

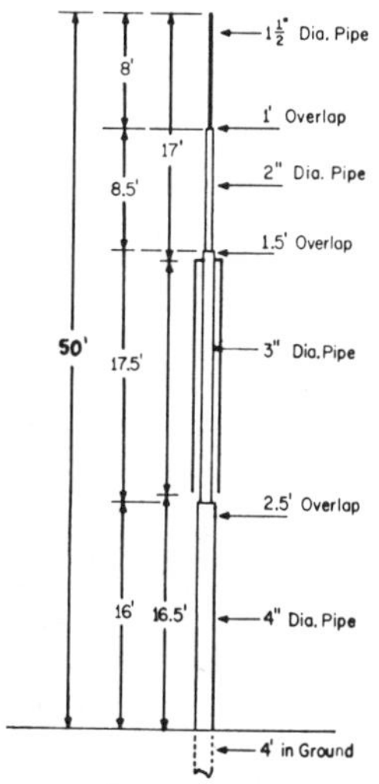

Fig. 136 — Dimensions of the antenna sections showing lengths, overlaps and location of the coaxial sleeve.

Coaxial Sleeve Construction

Now let us see how the coaxial sleeve is made. It is a simulated coax, or cage, with six peripherial wires as shown in fig. 139. It has a

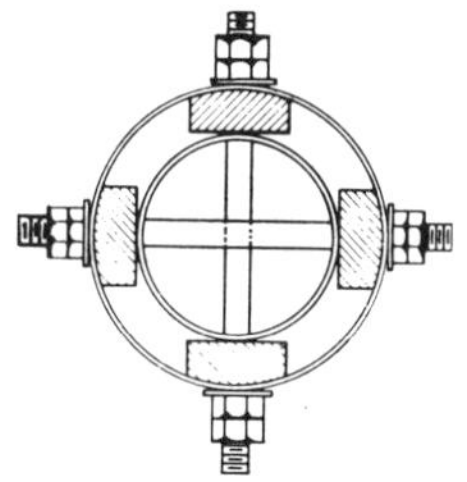

Fig. 137 — Cross section of the junctions showing construction and positions of the hardwood spacers. The hardwood blocks are wrapped with thin sheet aluminum. Bolt sizes for each junction are as follows:

Junction	Bolt Size	
4″ to 3″	(3) ½″ × 7″	(1) ½″ × 12″
3″ to 2″	(3) ⅜″ 0 5″	(1) ½″ × 12″
2″ to 1½″	(2) ¼″ × 4″	

9½″ diameter. The no. 9 aluminum clothesline wires are strung taut, with four spreader rings made of aluminum strap. The top and bottom rings are made of ¼ × 1″ strap, and the two middle rings are made of ⅛ × ¾″ strap. This material is of the Reynolds "do-it-yourself" variety available in many hardware stores, in 6 or 8 foot lengths. The brass ¼-20 nuts and bolts are also easily obtainable. A little trick is used in securing the wire under the nuts. Instead of turning the loop in the wire clockwise around the bolt under the nut, the loop is turned counter-clockwise. This brings the wire out on the right hand side of the bolt, and when tension is put on the wire it tends to turn

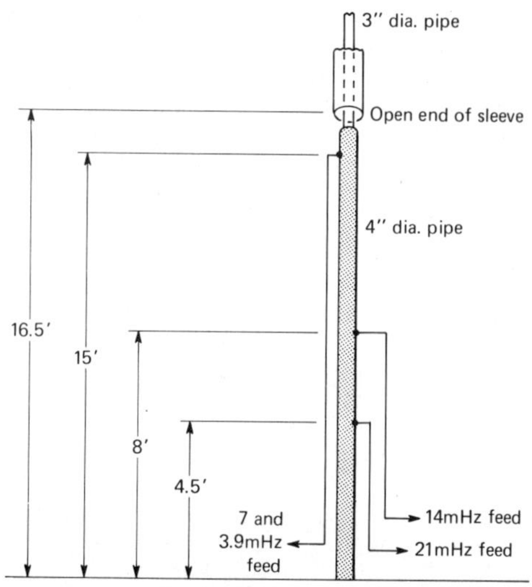

Fig. 138 — Location of the feed lines for each band.

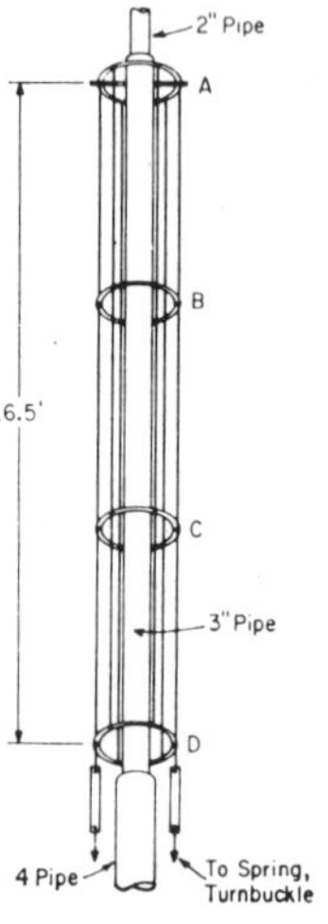

Fig. 139 — Construction details for the coaxial sleeve. Six lengths of #9 aluminum wire form the cage and are held in place on the aluminum rings by ¼-20 × ¾" brass screws and nuts. Rings A and D are made from ¼" × 1" aluminum with a 9" i.d. Rings B and C are made from ⅛ × ¾ aluminum with a 9¼" i.d. The two bottom brass bolts to which the 12" Johnson insulators (#136-112) are attached, measure ¼-20 × 1¼".

the nut clockwise and tighten it. This prevents the nuts from working loose. The open end of the sleeve is pulled off by means of a pair of heavy springs and turnbuckles, with an E. F. Johnson type 136-112 12" strain insulator in series with each. The top ring of the sleeve is held securely in place by double nuts on the 12" long ½" bolt which passes through the mast joint. The turnbuckles at the bottom are similarly fastened to the 12" long ½" bolt which passes through the lower mast junction. The sleeve is kept under considerable tension by means of the heavy springs, with the turnbuckles turned down to their limits.

Feed Line Construction

The connecting clamps for the upper ends of the feed rods are shown in fig. 140. There are three of these, made from ⅛" × ¾" aluminum strap, with ¼-20 brass hardware. After the final adjustment, double nuts are used to prevent the bolts from working loose. Figure 141 shows the design of the mounting ring for the bottom insulators of the feed rods. This ring fits on the mast about 18" above ground level. The insulators are of a surplus type, square in cross-section and 2½" long. I obtained them as part of some large s.p.d.t. knife switches which I bought from Barry Electronics, 512 Broadway, New York City. Each switch has three of these insulators, and I bought enough to give me a good stock of these handy items. Insulators made by Millen or E. F. Johnson could be substituted.

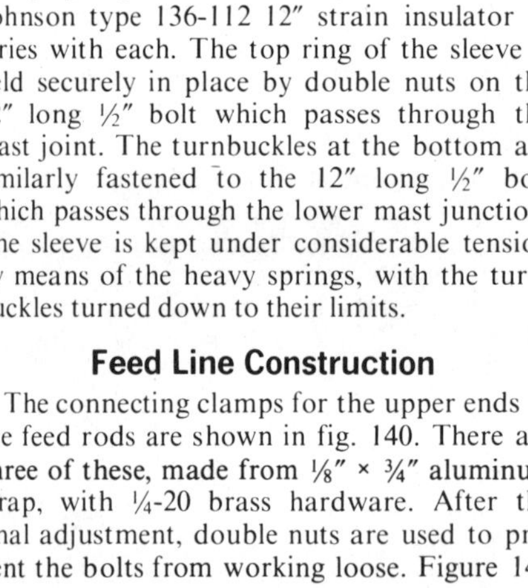

Fig. 140 — Construction details for the upper connecting clamps for the feedlines. They are fabricated from ⅛" × ¾" aluminum strap and held together by ¼-20 × 1" brass screws. Three straps are required.

The aluminum strap is flattened beneath each insulator, as shown in fig. 141, to give a smooth bearing surface for the insulator. There is a similar strap, made to mount only one insulator, which is used to brace the middle of the 13 foot 7/3.9 mHz feed rod to keep it from vibrating in a wind.

Figure 142 shows the detail of the heavy ground clamp used at the base of the mast. It is made of the ¼" × 1" strap, and is held tightly around the 4" mast by means of the ¼-20 bolts with double nuts. It is placed about 4" below ground level, and is used as the tie point for a piece of the ¼" × 1" strap which serves as the ground bus to the tuning unit and as the connecting point for the twelve buried ground radials. The 12 radials are of the No. 9 aluminum wire and are pushed into slots in the sod to a depth of about 3". Two of them are over 60 feet long and the others vary in length from 35 to 50 feet, fitting within the limits of the property. The coaxial feed cable and the relay control cable are buried from the house to the mast. The coaxial line is RG-8/U, and the control cable is plastic-covered electrical cable designed for direct burial. As much as the original Mark II setup as possible was used without change.

Assembly

This antenna mast was put together on the ground, the sleeve was fastened in place, all bolts tightened, and then it was raised easily by three people, and set in a 5" post hole drilled with a post hole borer. It was a bit awkward to handle because of its 50 foot length, but was not heavy, weighing no more than 30 to 35 pounds. It is quite strong in spite of its 50 foot height. During the 50 knot wind experienced shortly after its erection, the top of the mast bent over about 8 feet out of plumb, but when the wind died there was the mast, as straight as before. The movement of the sleeve in normal winds has no effect on loading.

Tuning Unit Adjustments

The operating sequence of the tuning unit relays is as follows:

Freq.	K_1	K_2	K
21 mHz	on	off	on
14 mHz	off	off	off
7 mHz	off	on	off
3.9 mHz	on	on	on

On 3.9 mHz I used a simple "L" network, with C3 in series with the + j110 ohms feed point reactance, adjusted to leave about + j20 ohms of X_1 in the series arm of the "L", with fixed capacitors C_6 and C_7 as the shunt arm. The input impedance is 50 + j0 ohms at the connection to the coaxial line. On 7 mHz the relays cut in an "Omega" match, consisting of two variable capacitors, C_4 and C_5. C_5 is the shunt arm of the "Omega" from feedpoint to ground, and C_4 is the series arm leading to the coaxial line. On 7 mHz the input impedance of the feed point is more favorable. Again, the input impedance is adjusted to 50 + j0 ohms. On 14 mHz, the connection point of that feed rod to the mast was adjusted, and C_2 was tuned to give an input impedance of 50 + j0 ohms.

On 21 mHz the same thing was done using the length of the 21 mHz feed rod and capacitor C_1 to obtain 50 + j0 ohms. Relay control was by means of switch S_1 on the operating desk as described in Chapter Fifteen. The back contacts on the relays are used to ground unused feed rods. This was especially necessary in the case of the 7/3.9 mHz feed rod, as a match could not be obtained on either 14 or 21 mHz with its floating.

One might get the impression from the preceding paragraph that a General Radio R. F. Bridge is necessary for tuning up. It certainly is a very handy tool, because it enables one to read accurately both the magnitude and sign of reactance and the magnitude of resistance, for each change in tuning. However, it is not an absolute necessity. One may use an s.w.r. meter, inserted in the coaxial line right at the input to the tuning unit. This must be an s.w.r. meter which will handle power, however, as adjustment of the variable capacitors is done with the transmitter on and feeding r.f. power to the antenna. This is really, a two-man job, with one person at the transmitter to turn it on and off for stub adjustments. He must also see that adjustments. being made at the tuning unit do no cause overload at the final amplifier. In my own case, I did not care whether the 4-1000A, running at low voltage and about 100 watts input, went out of tune or not, as long as I was able to get enough output for s.w.r. meter readings. The 4-1000A could not be hurt at all under such conditions. However, the proud

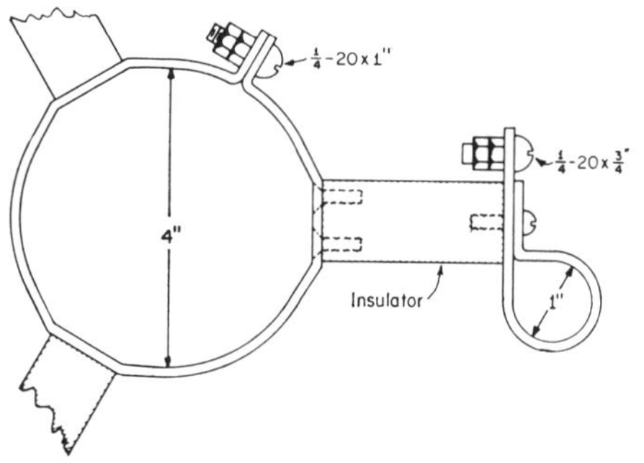

Fig. 141 — Base mounting ring for the 3 feed lines. It is constructed from ⅛ × ¾" aluminum strap. The area beneath the 1⅛" × 1⅛" × 2½" insulators are flat for solid support. The tubing clamps (3) are also fabricated from ⅛ × ¾" stock. The hardware is brass of the dimensions shown.

owner of a sensitive pair of 6146's would not be so fortunate, and should have someone standing by the transmitter to retune the final tank if required!

An appropriate type of s.w.r. meter is the "Beercan S.W.R. Indicator" described on page 96 of November 1957 *CQ*. It is designed to read s.w.r. under power.

Using the s.w.r. meter, adjustment will take longer than with an r.f. bridge, because it is a cut-and-try operation. At high values of s.w.r. it is quite difficult to detect the results of changes in tuning on a s.w.r. meter. For example, a reactance change from +j175 to + j150, while appreciable and in a favorable direction, would show very little change on the s.w.r. meter. Or, a change from +j50 to -j50 in reactance, with the *R* term remaining about the same, would not show up on the s.w.r. meter as a change unless one passed through the j0 point slowly in the tuning process. However, don't give up hope at this point. The job

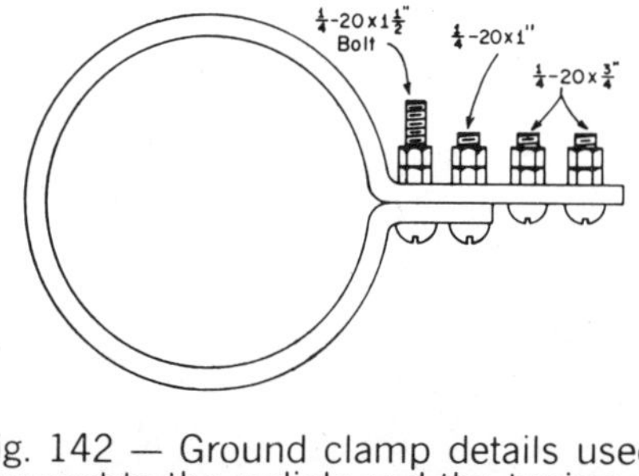

Fig. 142 — Ground clamp details used to connect to the radials and the tuning unit. It is made from ¼ × 1" aluminum stock and furnished with brass hardware.

Fig. 143 — Circuit of the tuning unit used to feed the Mark III Colinear vertical. The control circuit is connected to the remote tuning unit by underground plastic cable. S is shown in the 14 mHz position and all relays are off.

C_1,C_2,C_4—110 mmf vacuum variable.
C_3—750 mmf vacuum variable.
C_5—150 mmf Air variable.
C_6,C_7—0.0011 mf 5 kv micas. Sangamo type F-2 or equiv.
K_1,K_2,K_3—D.p.d.t. relay 117 v.a.c. coil. Ameco type 51 or equiv.

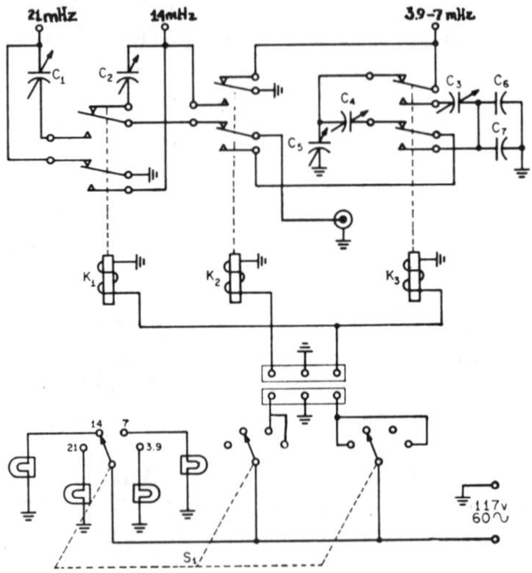

can be done with an s.w.r. meter and some patience and care.

In fact, a future tuning unit which I build will contain a built-in s.w.r. meter as a permanent fixture, with provision for shorting out the sensitive microammeter for its protection when not in use.

Performance

The Mark III provided many excellent DX contacts on 14 mHz, which it was designed to do, and it outperformed nearby Yagis in laying down strong signals in overseas places like Europe and Australia. On 7 MHz it gave very strong signals over the whole of the United States and Canada, depending on skip conditions, and provided many excellent DX contacts at night. It is useful in 3.9 mHz for short-range daytime contacts out to 500 or 750 miles, and at night it reaches out 2000 or more miles. Being able to use it on 21 mHz was an unexpected bonus. I had not originally so intended, and I have no idea of the current distribution nor vertical pattern shape on the 21 mHz band. ■

17

BROADBANDING THE MARK III ANTENNA ON 80 METERS

The Mark III, originally designed to cover the 3.8 to 4.0 mHz portion of the 80 meter band can be made to cover the entire 3.5 to 4.0 mHz. This chapter illustrates the use of the Smith Chart to devise the corrective network that broadbands the Mark III.

CHAPTER Sixteen described the Mark III DX Antenna. In the first three months after the appearance of that article in print in *CQ* I received over two hundred letters from readers inquiring about various aspects of antenna design. Many readers wished to be able to use the Mark III over the whole 80 meter band, and wondered how the s.w.r. over the whole band could be improved. It was originally designed to cover only the 3.8-4.0 mHz range. Some computation and a few neat tricks with that handy tool, the Smith Chart[58] have resulted in the information contained herein.

Use of the Smith Chart makes the solution of a problem such as broadbanding the Mark III 80 meter feed quite easy. The input impedance of the Mark III 80 meter feed versus frequency is tabulated below (tuned for 3.9 mHz):

Frequency	*Impedance*	*Normalized Value 50 ohm basis (R+jx)*
3.5	6.64+*j*5.6	.13+*j*.11
3.6	11.20+*j*7.66	.22+*j*.15
3.7	18.30+*j*9.69	.36+*j*.19
3.8	29.60+ *j*10.20	.57+*j*.20
3.9	51.00+*j*0	1.00+*j*0
4.0	58,80 -*j*33.10	1.15 -*j*.65

These normalized values of *R+jx* are plotted on the chart in fig. 144 as curve "A", giving a complete picture of impedance versus frequency in the range of interest. It may be seen that the 2.0:1 s.w.r. circle includes that portion of the curve between 3.8 and 4.0 mHz, but that the s.w.r. at 3.5 mHz is horrible, being about 7.5:1.

If the Mark III were tuned for 1:1 s.w.r. at the center of the band, 3.75 mHz, the curve would lie wholly within a 4:1 s.w.r. circle. The high s.w.r. at the extreme portions of the band is due to the *Q* of the feed point of the antenna, which is about 12. We can do something to improve this s.w.r. situation by adding several components to the matching network, as we shall see. The techniques used herein are employed in the design of shipboard antenna matching networks for Naval use, where it is required that an antenna be matched to a transmission line over a very wide frequency range, with s.w.r. of 2.5:1 or so.

The equivalent circuit of the Mark III feed, as tuned for 3.9 mHz, is shown in fig. 145A. As may be seen, it is a simple *L* network whereby

[58]Smith, P. H., "Transmission Line Calculator," *Electronics,* January 1939 Amis, P. C., "Antenna Impedance Matching-Part I." *CQ,* November 1963 page 63, Part II, December 1963, page 33.

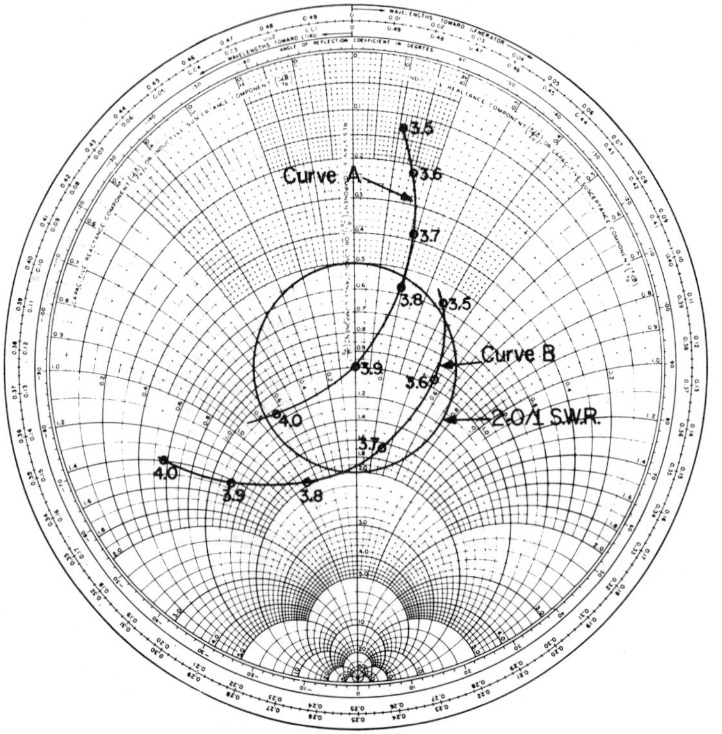

Fig. 144 — Curve A plotted on the Smith Chart above is with the Mark III antenna tuned to 3.9 mHz. The 2:1 s.w.r. circle includes only frequencies from 3.8 to 4.0 mHz. Curve B was developed with new values in the L network (fig. 146) and is positioned properly to be folded back on itself by the use of additional networks components.

the 9 ohm antenna feed point resistance R_a is matched to the 50 ohm line. The series inductive arm of the network is made of two parts as shown in fig. 145B. One of these is the positive reactance of the antenna feed point X_a, which is $+j113$ ohms. The other part is the negative reactance of the series capacitor, which at 3.9 mHz is $-j93.4$ ohms Thus the series arm is the sum of the two, or $+j19.6$ ohms. The shunt arm of the *L* network is $-j23.8$ ohms. (There are true values, not normalized.)

Mismatching the L Network

Our first step is to intentionally mismatch the *L* network in order to raise its input impedance, by a factor of 1.9, to a new *Z*, of 96.9 ohms. We also tune it for the center of the band, 3.75 mHz. The reasons for this will be

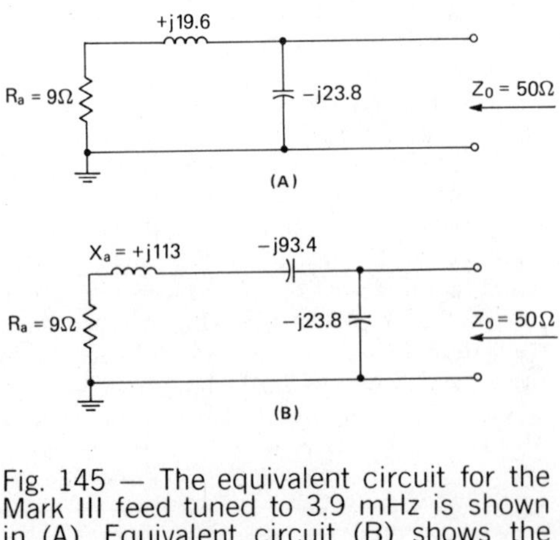

Fig. 145 — The equivalent circuit for the Mark III feed tuned to 3.9 mHz is shown in (A). Equivalent circuit (B) shows the negative reactance of the series capacitor which provides, in a series resultant, —**j**19.6 ohms.

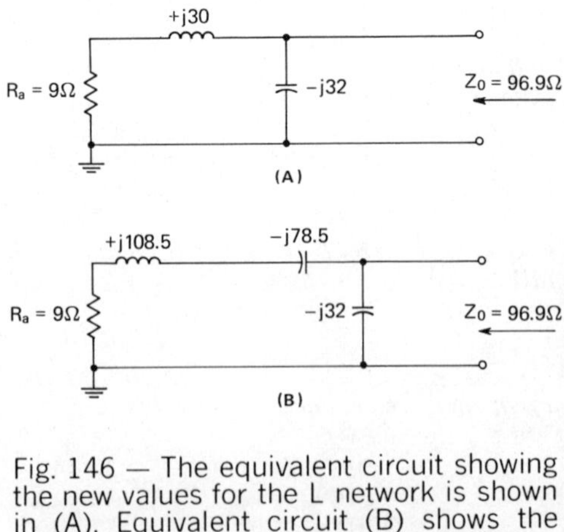

Fig. 146 — The equivalent circuit showing the new values for the L network is shown in (A). Equivalent circuit (B) shows the negative series reactance of —**j**78.5 thus giving the resultant +**j**30 as shown in (A).

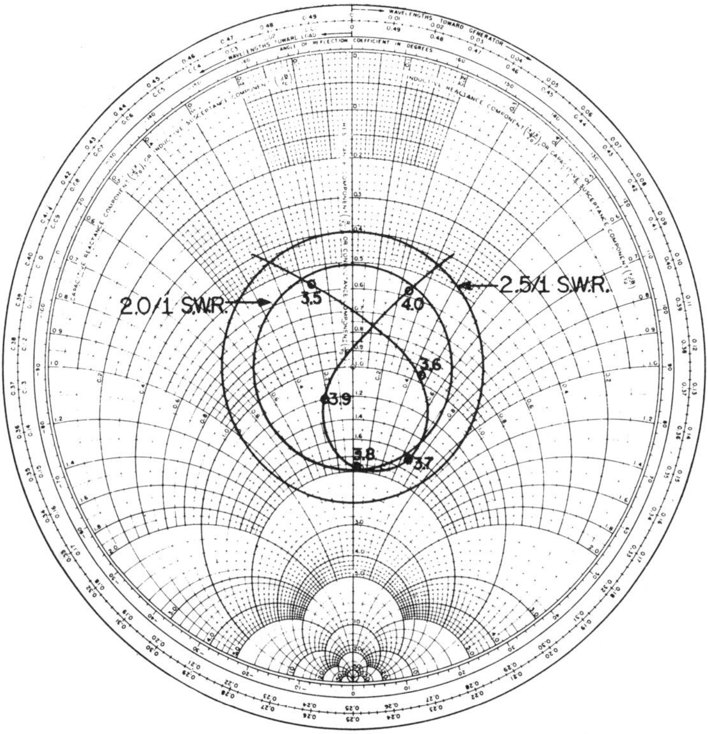

Fig. 147 — Curve B of fig. 144 has been bent back on itself as shown above and the 3.5 to 4.0 mHz portion lies completely within the 2:1 s.w.r. circle. While not perfect, it is usable over the entire 80 meter band and for those who wish to work outside the band (MARS) it is also usable.

seen later when the shape and location of the final input impedance curve are plotted and seen on the Smith Chart.

The new values for the *L* network of input Z_0 = 96.9 ohms at 3.75 mHz are shown in fig. 146A and 146B. The antenna feed point impedance is 9 +*j*108.5 at this frequency. Antenna resistance for all practical purposes remains at R_a = 9 ohms, whereas the reactance changes somewhat, with frequency. The negative reactance of the series capacitor must now be -*j*78.5 ohms, and that of the shunt capacitor must be -*j*32 ohms. The series arm of the *L* network is the sum of -*j*78.5 and +*j*108.5 ohms, or +*j*30 ohms, as shown in fig. 146A. The *L* network was computed from the formulae and charts in Terman's *Radio Engineer's Handbook*.

These new values for the *L* network components produce a new input impedance curve as frequency is varied. This is curve "B" in fig. 144. Note that the resistance components at 3.5 and 4.0 mHz are equal, at .57, and that the highest resistance component, 2.0, appears at about 3.77 mHz. The impedance values are also shown here in tabular form:

Frequency	*Normalized Impedance*
3.5	.57+*j*.38
3.6	.95+*j*.53

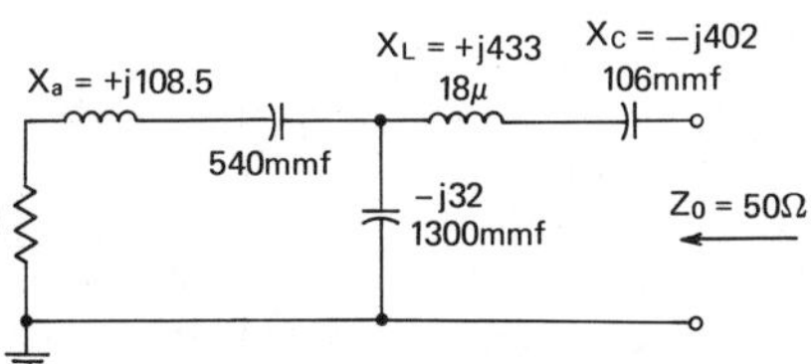

Fig. 148 — The new matching network shown above will permit the Mark III to operate over the 3.5 to 4.0 mHz band with an s.w.r. of less than 2:1 at any point. The true reactances and component values are shown.

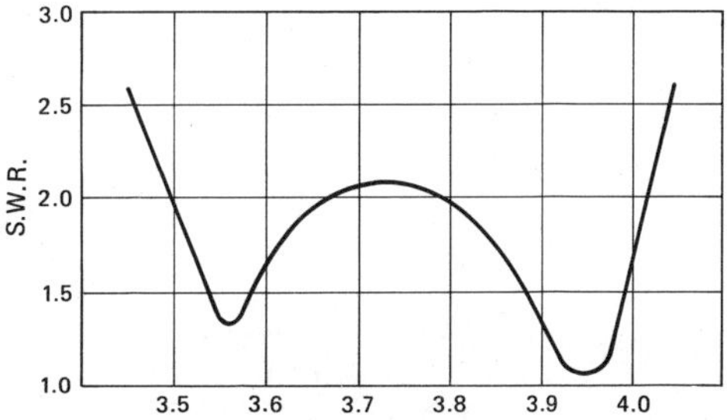

Fig. 149 — A rectangular plot of the s.w.r. of the modified Mark III.

3.7	1.67+*j*.30
3.8	1.96 -*j*.75
3.9	1.20 -*j*1.42
4.0	.57 -*j*1.39

Bending the Curve

We now have a curve which is properly placed on the Smith Chart to be amenable to being twisted back on itself like a pretzel by the insertion of a series differential network consisting of series positive and negative reactance, of such values as to cause their sum to be negative at 3.5 mHz and positive at 4.0 mHz. It is our purpose to cause the ends of the curve to move into the 2:1 s.w.r. circle, and to bring the whole curve as close to 1:1 s.w.r. as possible. We cannot make it perfect, but it will be usable over the whole 3.5-4.0 mHz band.

Inspection of curve B in fig. 144 shows that addition of -*j*.55 at 3.5 mHz and +*j*1.60 at 4.0 mHz will accomplish the desired result. The following set of simultaneous equations for the series network is used to compute the values of series X_l and X_c:

Where *a* is 4.0 mHz; $+j1.60 = +jX_{la} \ -jX_{ca}$, and *b* is 3.5 mHz; $-j.55 = +jX_{lb} \ -jX_{cb}$

Taking the ratio of frequencies 3.5/4.0 into account and solving, we find that X_{lb} is +7.90 and X_{cb} is -8.45. Computing them for all the frequencies of interest, we find the following:

Frequency	X_l	X_c	*Sum*
3.5	+7.90	-8.45	- .55
3.6	+8.12	-8.20	- .08
3.7	+8.38	-8.00	+ .38
3.75	+8.49	-7.89	+ .60
3.8	+8.60	-7.78	+ .82
3.9	+8.80	-7.60	+1.20
4.0	+9.03	-7.40	+1.63

Adding their sum to the values for curve B previously stated, we get:

Frequency	*Modified Z.*
3.5	.57 -*j*.17
3.6	.95+*j*.45
3.7	1.67+*j*.68
3.8	1.96+*j*.05
3.9	1.20 -*j*.22
4.0	.57+*j*.24

These new values of Z_0 are plotted in fig. 147, and one can see what has happened. The impedance curve is folded back on itself, and practically all of it lies wthin the 2:1 s.w.r. circle. Even the extremes at 3.5 and 4.0 mHz are not too bad for those who wish to go outside the band limits for MARS operation. Now you can see why we raised the input *Z* and balanced the curve about 3.75 mHz.

New Circuit

Figure 148 shows the new circuit diagram of the matching network with both true reactances and component values indicated. The true reactances are obtained by multiplying the normalized Z_0 values by 50 ohms, and the component values are based upon a center frequency of 3.75 mHz as stated above.

For those interested in voltage ratings of components, it should be stated that the series capacitors in the L and differential circuits should be rated for at least 2000 volts if a power of one kilowatt is contemplated. The shunt capacitor does not require such a high voltage rating, but it should be rated to carry at least 6 amperes at 4.0 mHz. I have had many inquiries about my use of variable vacuums. I used them because they were available to me, and they are very stable under all conditions. Air variables may be used in the Mark III tuning unit, provided they are in a completely weatherproof box.

In the above computations I have neglected losses in the coils and capacitors. This is perfectly legal if we use good quality components. To allow for stray capacities, however, we should make the components variable over a portion of their range. Their approximate values may be found by measuring with an impedance bridge. The final adjustment will have to be made by use of either a General Radio r.f. bridge to check the input impedance, or with an s.w.r. meter which can be operated under power.

Summary

This technique of altering the input impedance of a tuning network or an antenna and twisting the curve back on itself like a pretzel on the Smith Chart can be applied generally to other antennas and matching networks, with certain limitations. The input impedance of the load being altered must be of *positive reactance at the lower frequency,* and of *negative reactance at the higher frequency,* like that in curve *B* of fig. 144. The differential series network which folds up the curve over itself can only be applied if such is the case. (A network of series X_L and X_C always looks more capacitive as frequency is lowered, and more inductive as frequency is raised.) If the load impedance is not as stated above, it must first be reversed in sign by a suitable transformation network before the series differential reactances can be applied.

A rectangular plot of s.w.r. versus frequency

for the modified Mark III 80 meter feed is shown in fig. 149. At no time is the s.w.r. exactly 1:1, although it approaches it closely at several frequencies.

Inasmuch as the Mark III is already sufficiently broadened on 7 and 14 mHz, even enough to permit MARS operation on frequencies at some distance outside the bands, no changes have been made in the feed for those two bands. The only other change in the Mark III since the original article was the removal of the 21 mHz feed. The vertical pattern was not good for DX on that band. Removal of the 21 mHz feed caused the feed point for 14 mHz to move down the mast about 3 feet. This is the only change. The antenna performed well for 20 months and no further changes were made. ■

18

THE MARK IV DX ANTENNA

This six band antenna operates as a vertical on 160, 75 and 40 meters and as a beam on 20, 15 and 10 meters. The tower is a 40 foot aluminum ladder that does double duty as the beam support and the vertical radiator.

THOSE of you who read CQ regularly will recall that I am a strong proponent of vertical antennas for amateur use. In Chapter One of this book I discussed the ways in which a vertical antenna can be useful at h.f., and compared it both favorably and unfavorably with horizontal antennas and Yagi arrays. In past years I have been swamped with mail asking for information, and several articles by others appeared in CQ and other magazines both praising and condemning vertical antennas. However, in spite of those who condemn, verticals *do* work as stated. To use a statement which applies well in such cases, the operator listening to the signal doesn't know it is coming from an antenna which can't work, so he is able to hear it QSA5. It is the purpose of this chapter to show how some simple principles of antenna design can be applied to a very common situation which exists in the backyards of many amateurs.

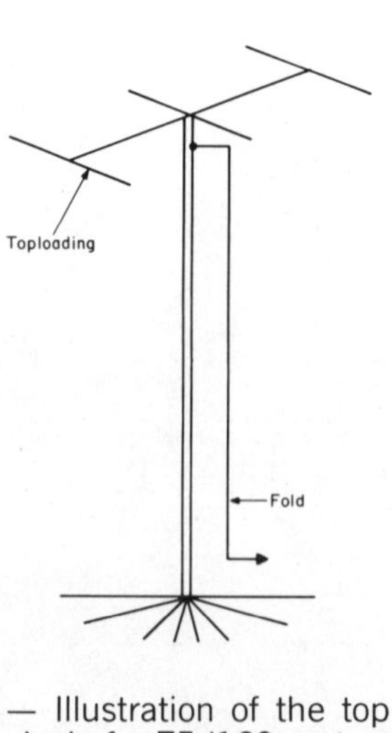

Fig. 150 — Illustration of the top loaded unipole for 75/160 meters.

In July of 1966 it appeared that we would be moving our household overseas from Maryland. I wanted to make a change in my antenna situation simultaneously with that move. First, I wanted an antenna for 10 meters. I had never installed one simply because of the sunspot cycle which had made the band quite useless for DX work for the previous six or seven years. Second, I wanted an antenna which would discriminate against QRM from unwanted directions. This is one of the faults of the vertical antenna when working DX on 10, 15 or 20 meters, and I pointed this out in Chapter One. Third, I wanted an antenna which would be simple to erect and dismantle, and which would cover as many bands as possible with one structure, including the 160 meter band. Fourth, simultaneously and in

View of the tuning set up used to obtain the impedances shown in figs. 153, 154, 155 and 156. The equipment consists of an LM signal generator, a 916A General Radio Bridge and an ARC 5 receiver used as a detector. The 4″ × 4″ supports can be seen with the 2″ pipe running through the bottom rung of the ladder.

common with the above factors, I wanted to "buy" some forward gain on 10, 15 and 20 meters.

An ad in our local Maryland/D.C. area magazine *AutoCall* caught my eye. A local amateur was selling his HyGain TH3 Mk2 Thunderbird beam plus a CDR TR-44 rotator for half price! An idea formed in my mind immediately, and I bought the beam and the rotator, and stored them in the basement pending further study. Several of my friends heard of my purchase, and said, "You must be sick!" To that I replied, "No, I'm not — I'm going to make a top-loaded vertical radiator out of it." Some then thought I was completely mad, but others saw a glimmer of hope and light and decided to await further developments with interest. Meanwhile, I went on working stations with the Mark III, stations who didn't know it could not work, on 20, 40 and 75 meters. There was a separate ⅝ wave whip for 15 meters.

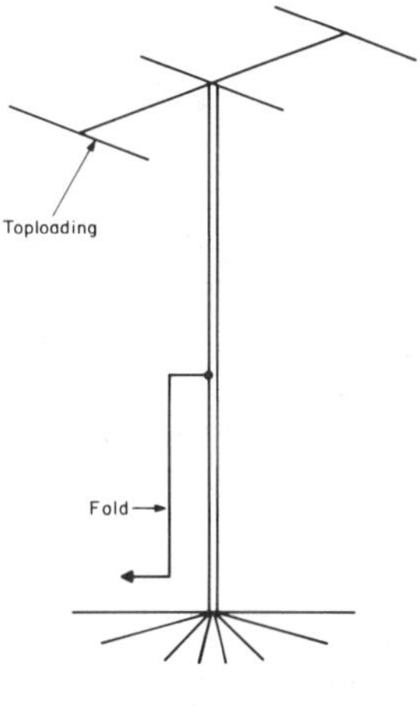

Fig. 151 — The 40 meter unipole feed is made of 1″ aluminum tubing tapped 20 feet up the mast.

View of the upper 20 feet of the tower. The coaxial line may be seen on the right side going up the tower on standoffs. The 40 meter feed wire, since replaced with one inch tubing, is at the left. The 2" cross-members for the beam support and at the guy levels can be seen in this view.

The Ladder Vertical

Having bought an excellent beam and rotator at a bargain, I wanted to do the same thing regarding a tower, bearing in mind that I wanted something easy to erect, dismantle, transport and re-erect. The mailman brought a sale "flyer" from Montgomery Ward. In it were listed aluminum ladders. I picked out a two-piece 40 foot type, and ordered it for the sum of $59.95 plus delivery charges. The 40 foot ladder would extend to 36 feet, and would be a very rigid support for the beam, when guyed. It would collapse to 20 feet, which would fit inside a moving van. It would be easy to erect, and easy to take down. It would also serve as a vertical radiator. It could be climbed with great ease. There would be a minimum of erection hardware.

View of the tuning unit, closed up. The 40 meter feed comes out the insulator on the left side, and the 75/160 meter feed (coaxial line to the beam) comes out the insulator on the right side. The line actually goes through the hole in the insulator. The box is water-tight, and locked.

Thus was the idea of the Mark IV six-band antenna born. The HyGain antenna would function as a beam on 10, 15 and 20 meters, at a 40 foot height which would be a fair compromise for all three bands as stated in Chapter One. The tower with the beam as top-loading would function as an efficient vertical radiator on 40, 75 and 60 meters. I especially wanted to be able to use it on 160 meters, because my s.s.b. exciter[59] covers that band. I had a long wire for use on Navy MARS frequencies, but use of the vertical on 160 meters was very desirable. With the purchase of guy wire, guy insulators, and miscellaneous hardware such as thimbles, clips, eye bolts and turn-buckles, the package was completed within several weeks, and ready for erection.

That job was done in one weekend, during which two other local amateurs helped me lift the Mark III out of its hole in the ground, and lay it down tenderly. It was given to one of them with grateful thanks for his help. Next came the erection of the new antenna. The rotator and beam were placed in position, with the ladder hinged at its bottom and lying on its side. The ladder was then pushed upright, and the bottom set of guys were fastened to their anchors. The ladder was plumbed by adjustment of guy tensions. The upper section of the ladder was then pulled up, the two upper sets of guys were fastened to the anchors and adjusted until the ladder was plumb all the way up to the top, and the coaxial line, rotor control cable, and the 75/160 meter feed and the 40 meter feed were put in position. The erection job was complete.

The beam was then tested on 10, 15 and 20 meters, with its coaxial line connected directly to the line into the station. The s.w.r. forward gain, and front-to-back ratio were exactly in accordance with the manufacturer's specifications, which pleased me very much. The dimensions of all elements of the beam had been previously set for the phone bands in accordance with Hy-Gain's instructions, and I had left it

[59]Lee, P. H., "The Ultimate SSB Exciter," *CQ*, Feb., 1967, p. 24

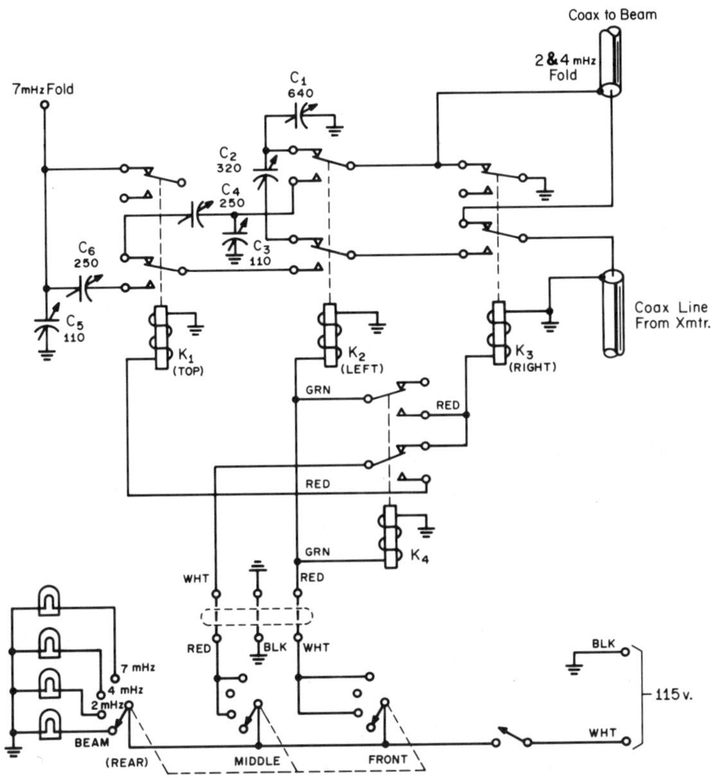

Fig. 152 — Circuit of the tuning unit used to feed the six band Mark IV antenna.

that way. I did not use a balun with the beam, but wound a 12 turn coil of RG-8/U coaxial line 6 inches in diameter, in accordance with Hy-Gain's suggestion, and used it as an isolation choke. It functioned perfectly.

The next step was the rebuilding of the Mark III tuning unit to fit the new vertical radiator for 40, 75 and 160 meters. This was accomplished in another weekend, and the whole thing was then usable on 10, 15, 20, 40, 75 and 160 meters, with bandswitching from the control position in the station.

Unipole

That concludes the narrative — now for the technical details. First, the operation of the beam itself is exactly in accordance with Yagi theory, and it will not be covered here. It functions very well, and certainly speaks well for the design and manufacturer. But how do we cover 40, 75 and 160 meters with this structure? Well, we treat the structure as a folded unipole.[60] [61] [62]

You are all familiar with folded dipoles, I am sure. It is one of the most common antennas in use today, for the "driven element" of many TV receiving arrays in a folded dipole. If you mount a folded dipole vertically, cut off the bottom half of it, and replace that with a ground plane, you would have a folded unipole. If you add some top-loading[63] [64] [65] in the form of some radial wires, a top-hat, or the long horizontal boom and elements of a 20 meter Yagi array, you will thereby greatly increase the electrical height of a vertical radiator over its actual physical height.

Figure 150 shows the schematic of the top-

[60]Leonhard *et al,* "Folded Unipole Antennas," *IRE Transactions on Antennas and Propagation,"* July, 1955.
[61]Lewis, "Use of Folded Monopoles in Antenna Arrays," *op. cit.*
[62]Mullaney, "The Folded Unipole Antenna for Broadcast," *Broadcast Engineering,* Jan. 1960.
[63]Smith and Johnson, "Performance of Short Antennas," *Proc. of the IRE,* Jan. 1960.
[64]Monser and Sabin, "Antenna Design for Maximum LF Radiation," *Electronics,* June 3, 1960.
[65]Smith *et al,* "Very High-Power Long Wave Broadcasting Station," *Proc. of the IRE,* August, 1954.

Interior view of the tuning unit. The air variable capacitors $\mathbf{C}_1$ and $\mathbf{C}_2$ are at the lower right. The others are all vacuum variables. Relay **K** is in the upper left; $\mathbf{K}_2$ and $\mathbf{K}_3$ are at the bottom along with the extra control circuit relay. All r.f. wiring is done with #12 enameled copper wire. The 75-160 meter feed (coax to the beam) enters on the right and the 40 meter feed is visible on the left.

loaded folded unipole for 75 and 160 meters, as it is configured in this installation. Note that the feed wire or "fold" goes to the top of the vertical radiator. With the radiator grounded at its base, the impedance looking into the fold was about 2.0 + *j*140 ohms at 1.812 mHz and about 35 + *j*370 ohms at 3.900 mHz.

At 7.250 mHz the impedance looking into this fold was out of range of the General Radio 916A r.f. bridge. Therefor, a separate fold was used for this frequency, and it was tapped on the radiator at a point 20 feet above ground. The impedance looking into this fold, made first of #9 aluminum wire, was 18 + *j*480 ohms. Changing the fold to 1 inch aluminum tubing lowered the impedance to 36.5 + *j*355 ohms. Figure 151 shows the schematic for the 40 meter feed.

160 Meter Fold

For the 75 and 160 meter fold, a separate wire was not used. The outer braid of the RG-8/U coaxial line which feeds the beam was lifted off ground, and was used as the fold wire

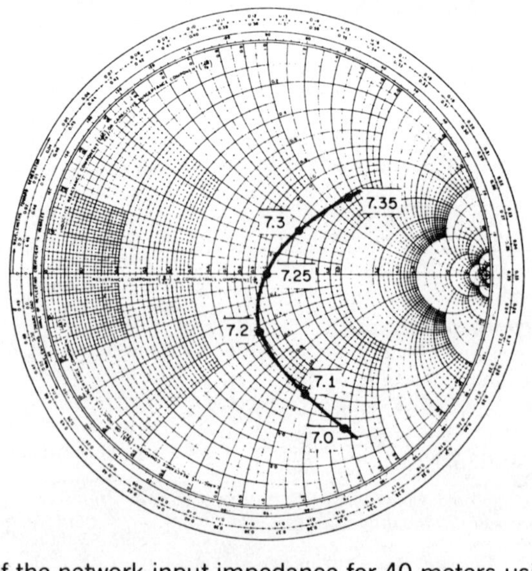

Fig. 153 — Plot of the network input impedance for 40 meters using a #9 wire for the fold.

to feed the radiator as a folded unipole. The RG-8/U outer braid is permanently connected to the top of the tower, but the RG-8/U came down the side of the tower on 3 inch standoff insulators. Thus, by ungrounding its bottom end and feeding it as a fold wire, the folded unipole feed for 75 and 160 meters was accomplished. The tower and the coaxial line don't know the difference, and this type of feed arrangement works very well. The 1 inch tubing fold is used for 40 meters, and it is supported on the opposite side of the tower on 4 inch standoff insulators.

Between the top of the tower and the toploading formed by the beam elements there is the rotator motor and housing. I checked with representatives of CDR regarding inserting a rotator in an active r.f. element in this manner, and they saw no problems, because of the many internal contacts through the ball bearings in the raceways between the fixed and rotary shells. I did bypass all rotor control cable leads to ground with 0.01 mf ceramic bypass capacitors at the base of the tower, however, to prevent any possibility of r.f. being carried back into the house on the rotor cable.

The tuning unit was adjusted with the r.f. bridge with the beam elements parallel to the plane of the ladder. However, I subsequently noted no change in transmitter loading nor in s.w.r. when I rotated the beam while the transmitter was operating on 40, 75 or 160 meters. I did not really expect much change.

Tuning Unit

The schematic of the tuning unit is shown in fig. 152. I used as much as possible of the Mark III unit, changing very little of the internal wiring. An extra relay was inserted, however, to provide a means for controlling the three r.f. relays with only two wires and one neutral in the buried cable. I did not care for the idea of pulling the cable out of the ground and replacing it. The extra relay was a much more expeditious solution.

The RG-8/U coaxial cable went up the tower through the hole of an E. F. Johnson type 135-67 standoff insulator which served as a feedthrough insulator. The hole was the exact size, and a bit of caulking compound was used to insure watertight integrity. This coaxial line must be insulated from ground so that it can be ungrounded and fed as the fold on 75 and 160 meters. Capacitors C_1 and C_2 form an L network for the 160 meter tuning. As shown in the table associated with fig. 152, relay K_3 is "on" for 160 meters. The shunt capacitor, C_1 is used to partially resonate the tuned circuit formed by the fold and the tower. Capacitor C_2 cancels out the residual positive reactance, and enables a perfect resistive match to the coaxial line from the transmitter.

For 75 meters, relays K_2 and K_3 are "on," and C_3 and C_4 form a similar L network for

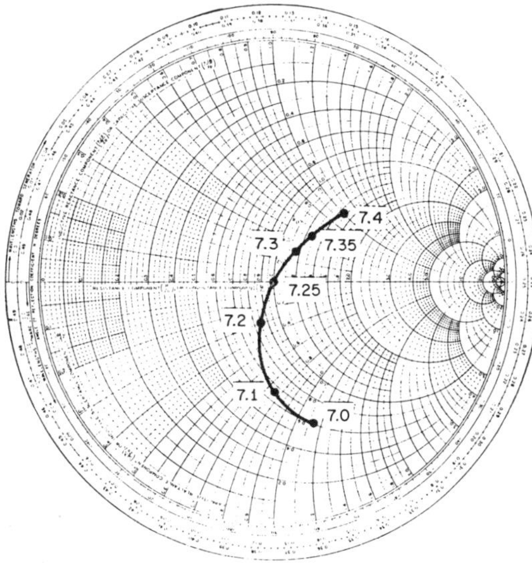

Fig. 154 — Plot of the network input impedance for 40 meters using one inch aluminum tubing for the fold. Note the increased bandwidth in the 2:1 s.w.r. circle.

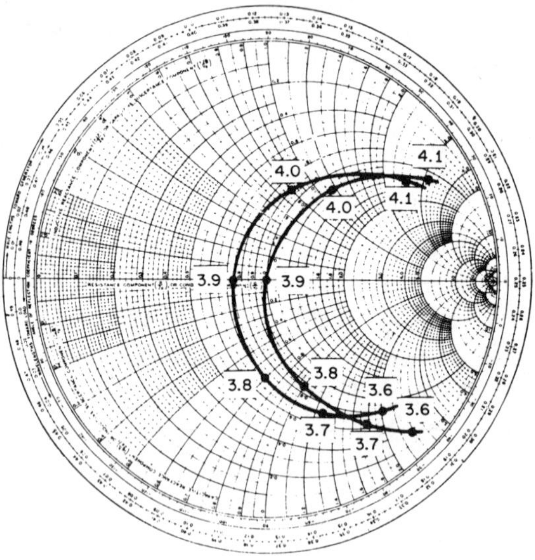

Fig. 155 — Plot of the network input impedance for 75 meters.

this band.

On 40 meters, all three relays K_1. K_2 and K_3 are "on," and the feed is transferred to the 40 meter fold, where C_5 and C_6 form the L network

Capacitors C_1 and C_2 are a pair of large transmitting air variables obtained from Barry Electronics in New York. Substitutes may be used but about 500 mmf is actually required in C_1, and about 120 mmf is required in C_2, in this particular installation. Capacitors C_6. C_3, C_4 and C_5 are small variable vacuum capacitors, but air variables can be used here if they have at least 2500 volts rating to handle 2 kw p.e.p.

On 10, 15 and 20 meters, relays K_1, K_2 and K_3 are all "off," and the power is fed directly through from the transmitter coaxial line to the beam via the upper contacts of relay K_3. Breaking the coaxial line and inserting relay K_3 caused a very slight amount of s.w.r. on 15 and 10 meters, but this is not s.w.r. in the beam. It is merely s.w.r. which exists on the line from this *relay* back to the transmitter. I know of no way to get rid of it, because this switching cannot be done with a coaxial relay, for the outer braid of the line to the beam must be lifted off ground for the 75 and 160 meter feed. Therefor I do not worry about the fact that the s.w.r. is 1.5:1 instead of 1.0:1, for example, on the transmitter side of the relay. *Que sera, sera.*

Impedance Plots

Impedance plots of the "L" network input impedances were made with the General Radio 916A r.f. bridge for 40, 75 and 160 meters. They are shown in figs. 153, 154, 155 and 156. The effects of the #9 wire *vs* the 1 inch tubing for the 40 meter fold are shown in figs. 153 and 154. It is obvious why I chose to use the tubing, for the bandwidth within the 2:1 s.w.r. circle is greater.

Figure 155 shows the effect of setting the 75 meter L network to produce an input impedance of 40 + j0 ohms versus an input impedance of 50 + j0 ohms at 3.9 mHz. With the former, the s.w.r. is never 1.0:1, but it is better than 2.5:1 over the limits of the 75 meter phone band. The s.w.r. across the band is better with this impedance setting than with a setting of 50 + j0 ohms. The same thing may be said for fig. 156 which shows the impedance plots of the 160 meter "L" network when set for an input impedance of 30 + j0 ohms *vs* one of 50 + j0 ohms at 1.812 mHz. With the former, the s.w.r. is never 1.0:1 across the band, but it is only slightly over 3.0:1 at 1.800 and 1.825 mHz. However, when the network is set at 50 + j0 ohms, the s.w.r. at the band limits is about 5.0:1.

Radials

The 16 ground radials of the Mark III were first used with this antenna on 40, 75 and 160 meters. However, I later discovered much corrosion of the No. 9 aluminum wire, and as stated in Chapter Twelve I put down 50 new radials of No. 16 copper wire, from 35 to 80

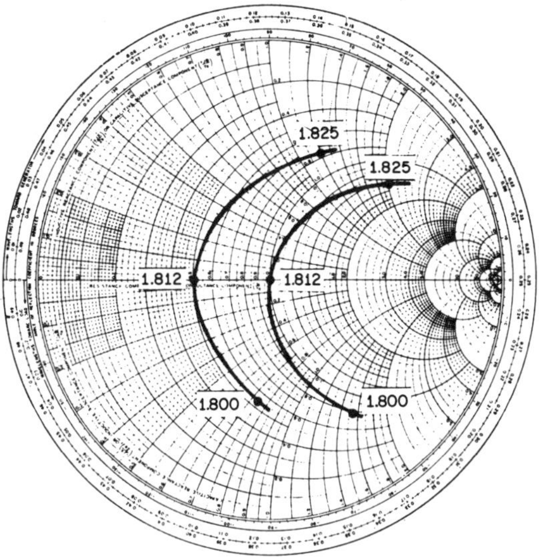

Fig. 156 — Plot of the network input impedance for 160 meters.

feet long. Connection was made to the tuning unit and to the base of the tower. Also, a 6 foot ground rod was driven into the earth at the base of the tower, to function mainly as a static and lightning drain.

Tower Rigging

There is nothing special about the rigging of the tower, except that at points of guy attachment to the ladder, no reliance is placed on the rungs of the ladder to take the lateral strain. A length of 2 inch aluminum tubing, with ends flattened and bent over, is securely bolted in at each such level and two pieces are used as supports for the pipe holding the beam rotator. The ⅜ inch eye bolts to which the guys are attached are actually bolted through the flattened ends of the 2 inch tubing, making a very strong guy attachment.

In addition to being held by "U" bolts to two of these 2 inch crossmembers, the rotator pipe is bolted to one of them by a 5/16 inch through-bolt which prevents the rotator pipe from twisting. The rotator is similarly secured to its upper and lower pipe sections, so that no twisting can occur.

One of the illustrations shows the spring steel clips by means of which the coaxial line was secured to the ends of the 3 inch standoff insulators. These were found in a hardware store, and apparently were intended to fasten tools to a tool board. There was also one such clip under the screw which held the inboard end of each insulator, and the rotator cable was slipped into these on the inside of the ladder leg. In this way both the downcoming coaxial line and the rotator cable were very easily detached and left hanging when it was desired to lower the upper section of the ladder.

Also, there was a coaxial connection using two PL-259 plugs and one PL-258 connector in the line just outside the feedthrough insulator, so that the coaxial line could be detached from the tuning unit. This junction was taped with several layers of waterproof vinyl tape.

The connection of the coaxial line to the top of the tower is made with two PL-259 plugs and one PL-275 adapter. The PL-275 adapter is fastened to the tower by a ⅛ inch aluminum plate. This junction is also well taped with waterproof vinyl tape and enables the beam to be detached and removed from the tower if desired, without removing all of the coaxial line from the tower. In doing such a waterproof tape job, one begins winding at the bottom of the junction, and spirals upward. Thus each layer is on top of and outside of the one below. A coat of quick-drying varnish is then applied.

The rotor cable bypass capacitors are contained in a small waterproof can which was mounted on one of the 4 × 4 ladder supports. One lead of the cable is *ground*, and the other six are bypassed to it, and it is connected here to the ground bus.

Tower Support

The 4 × 4 supports were amply strong to

held the weight of the structure, being imbedded in 4 feet of concrete. A 1 inch iron pipe through the bottom rung of the ladder served as a hinge, for laying the ladder over on its side.

The guy cables are of fourteen strand wire rope. Each one has its own turnbuckle for adjusting tension. Each cable is insulated from the tower and is also broken up into sections of less than 12 feet by small egg or "johnny ball," insulators. I have noted no change in s.w.r. on 10, 15 or 20 meters when I rotate the beam, which speaks well for its isolation from surrounding objects.

Results

The results on all bands have been outstanding. The beam performs exactly in accordance with the manufacturer's specifications, and I have had many excellent DX contacts with it. Apparently its height is a good compromise for all three bands, 10, 15 and 20 meters, although it is not as high as it should be for good low angle radiation on 20 meters. It performs best to date on 15 meters, and in Maryland it brought me many reports of being the strongest signal on the band from Europe, Africa, New Zealand and Australia.

Good DX openings have not been very plentiful on 10 meters, and I have not had so much opportunity to evaluate it on that band as I have on 15 meters. However, those contacts I have made with overseas stations on 10 meters have resulted in excellent reports.

The beam top-loads the 40 foot radiator so that it is quarter wave resonant at about 3350 kHz. It is thus a bit over ⅛ wave high at 160 meters, and about ⅝ wave high at 40 meters. These are excellent heights for the bands involved, and vertical patterns are produced which are optimum for the propagation on those bands, with good radiation efficiency.

On 160 meters I worked stations up and down the east coast, and in the midwest, with the exciter[59] "barefoot" and reduced to 50 watts at night. On both 40 and 75 meters there were excellent contacts out to 2000 miles. Full power of 1 kw voice (2 kw p.e.p.) is used on all bands except 160 meters.

Moving The Antenna

I built this antenna to be with me for a long time to come. We did not move from Maryland to overseas in 1967, but remained in Maryland until December 1970, when we moved to California. As planned, I dismantled the Mark IV Antenna, loaded it into the huge moving van that transported our household goods, and re-erected it here at the K6TS QTH in Thousand Oaks. The re-erection was easy, and was done with assistance from three of the local club members. In the new installation I made some changes, however. Instead of mounting the ladder on two 4 × 4 wood posts, I made a pipe framework in the form of an inverted letter "U" with the base of the "U" through the bottom rung of the ladder, and with the ends set in about 30″ of concrete. Those 30″ came very hard here in this QTH, for our soil is like adobe when dry! The base of this antenna will never move, even in an earthquake! The guy anchors which I had carefully unscrewed from the earth in Maryland and brought along in the van were used again here, but the soil was so hard that it was necessary to use the hose to soften up a hole into which to screw the anchors. Ever removing them will be a task to which I would not look forward to with enthusiasm.

The RG-8/U coaxial feed up the side of the ladder on standoffs was not used again as the feed for the 75 and 160 meter unipole. Instead the coaxial line was fastened to the inside of the tower leg and used merely to feed the beam. A No. 9 aluminum wire was substituted as the feed for 75/160 meters, giving me a slightly higher input resistance to the "fold" and a better Q, for the 160 meter feed. I have not plotted the Smith Chart for the input resistance of this No. 9 wire "fold" on 75 or 160 meters, but it is working fine. The 40 meter tubing feed was used again.

In leaving Maryland I pulled up the 50 ground radials from the back-yard lawn (the weather was very wet and the ground very soft at the time) and brought them along. They were re-installed here, but some had to be shortened, and now they are from 30-50 feet in length. Again, getting them into the ground was difficult, because the soil when dry is like adobe, and when wet it is like a good grade of glue! This area was ocean bottom in ages not too long ago, and the soil is very peculiar. Anyone reading this who lives here will know what I mean! The job was done with a lawn edger blade and a lot of "foot power."

In this installation the ground radials have been soldered to a 2″ wide copper strap which is bolted to the bottom of the ladder. The strap also connects to the bottom of the tuning unit box. The tuning unit itself is mounted on the side of the ladder about 1 foot off the surface of the earth. As before, the control cables for relays and rotator and the coaxial feed line are buried from the tower base into the house. Also, since the power limit on 160 meters is now higher than it was, I replaced the air variable capacitors with vacuum variables. With the power now allowed in daylight hours on 160 at a kw and 200 watts at night, the air variables arced at 2 kw p.e.p. and hence the change to facilitate higher power.

The tower is not as high as some in my immediate vicinity, but the Mark IV has given very good performance. There have been several contacts with the east coast on 160 meters, and many on 75 and 40 meters with this antenna from the Thousand Oaks QTH. And of course on 15 and 20 the Pacific Islands which were once DX from the Maryland location are now "in our front yard." Unfortunately the sunspot cycle has continued its downward trend, and contacts with Europe, Africa and the Atlantic have become more and more difficult since this move, and I have not really had the chance to evaluate the performance of the antenna here to get a comparison with the Maryland location, especially when working through the QRM generated to the middle and eastern United States. Long path contacts to South Africa, when the propagation has been open, have been good, as have contacts with VK, ZL, KH, JA, and all other Pacific locations. The antenna has performed well in working southeast to South America and the Caribbean, also. I am looking forward to the buildup of the sunspot cycle again, to really give it a fair test against some of the monster beams we have out here.

I think now that the only thing I will do to this antenna in the future is to use a taller, self-supporting crank-up tower in place of the ladder. I would like to get at least 60 feet of height here at the K6TS QTH, to get above a power line which runs to the rear of the property. The crank-up tower would be left cranked up, with jumpers connected across the junctions between tower sections. The 160 meter feed would go all the way to the top, the 75 meter feed would go up to about the 35 foot level, and the 40 meter feed would be connected at about 20 feet. At least, these locations would be a good start for the final adjustments which would again be done with the General Radio 916A r.f. bridge. This will be the future Mark V, but for now I am content with the Mark IV.

In conclusion I wish to express my sincere appreciation to my friend and associate, John H. Mullaney, W3NGJ[62] Consulting Radio Engineer, Potomac, Md., for his interest, ideas and help in connection with this project. ■

19

USE OF ANTENNA MODELS IN DESIGN WORK

YOU may well ask how the design engineer knows that an antenna performs in a certain manner. How are the patterns of the antennas described in this book determined? Obviously one does not want to build a full-scale v.l.f., l.f. or m.f. antenna for experimental purposes. Tower construction is very costly, as is the cost of ground systems. Even copper wire and cable is a very dear and scarce commodity these days. How then do we know that an antenna of certain height will do thus and so? Or that a certain top-loading scheme will do what one wants it to do? Or that an h.f. array will have a certain pattern?

As shown in Chapters Five, Nine and Ten, the patterns of m.f. directional arrays consisting of vertical elements can be computed, from the design information consisting of location of the elements in relation to each other, their phasing, and their current ratios. Patterns can be computed for the horizontal plane only, or for any desired vertical angle above the horizontal. This computation was done in past years manually, by use of log and trig tables, slide rule, charts, etc. and was quite a long and laborious process. The Tables in these Chapters were made up that way by myself as examples of the process. Several of the consultants designed analog computers which would do the job, actually tracing a pattern on polar paper as it was being computed. This was a major step forward in its day, and was quite a time-saver during the design process. It had to be followed by exact mathematical computations to satisfy the F.C.C., however. Now this type of work is done by digital computers which are programmed to do it. Such services are available to consultants and design engineers for a nominal fee for the computer time.

With regard to the v.l.f., l.f., or m.f. antennas consisting of a single tower, they are modelled in small scale quite accurately, and the models are tested on large ground planes consisting of copper wire, sheet, or screen. The scale is such as to bring the design frequency up into a range where measuring equipment is available. Effects of tower height, L/D ratio, top-loading, and guy configuration and feed arrangements can be determined quite readily by changing these parameters on the model, whereas at full scale any such changes would be quite prohibitive as to cost.

H.f. antennas can also be modeled at small scale by reducing their size by a factor which again will place the design frequency into a range of v.h.f. or u.h.f. where measuring equipment is available. It is quite common to test h.f. antennas on a v.h.f. or u.h.f. range. Effects of proximity of ground plane, element configuration, and nearby towers or objects can be determined experimentally.

Of course v.h.f. and u.h.f. antennas are already of relatively small dimensions, and they are usually modeled at their normal size, or in the case of an array which consists of many elements which would be unwieldy, at ½ or ⅓ size. Again the design engineer has to be careful to pick a scale for which measuring equipment is available at reasonable cost.

Accuracy of Results

How accurate are the results from these processes? In the case of m.f. vertical arrays, there have been many thousands of such arrays built for the broadcasters, the results have been made known in reports of proof-of-performance filed with the F.C.C., and the accuracy of the design can be proven beyond question. The formulas are well-known, and have been used for many years. Accuracy of design is assured. Ground conductivity and its effects have also been well-proven over many years, and the conductivity map shown in fig. 16 in Chapter Two is quite accurate. We now have conductivity maps of Canada, Mexico and the rest of North America, and these are used in groundwave computations in connection with problems involving antenna design where signal contours across the national borders must be known. These matters have evolved over the fifty years or so of m.f. broadcasting.

With regard to v.l.f., l.f. or m.f. antennas consisting of single towers, the model techniques involved in scale models are fairly accurate. Such studies have been made by the U.S. Navy and its contractors in the design of large antennas for those frequency ranges, and the results have correlated quite well between scale model and full-size antenna. One of the things which cannot be modelled at such reduced scales is conductor size and its effects. Skin effects in cables, for example, are quite different at v.l.f. than they are at h.f. A multi-strand 1″ cable is hard to model at a scale of 50 or 100 to 1, for example! The design engineer must be content with simulating such things as closely as possible.

Another thing which cannot be modeled is the effect of ground conductivity. Reference to Chapter Twelve will show why this is so. Depth of penetration of ground currents differs with frequency. For this reason, most model ranges which are used for design of v.l.f., l.f. or m.f. tower antennas employ copper ground planes consisting of radials, screen, or sheet and the patterns and other data which result are for the case of as nearly perfect ground as the designer can make it. Obviously then, he must translate his data into a form which takes into account a poorer ground than optimum.

H.f. arrays can be done full scale, but it is common to use a v.h.f. or u.h.f. range for modelling such antennas. Free space can be approximated by keeping the models at good height above ground and surrounding objects, but here again such results are more theoretical than actual, for free space conditions can never be achieved at h.f. Ground is always there, and its conductivity varies with frequency. The processes described in Chapter Twelve have to be used to take into account the presence of earth below a h.f. antenna.

V.h.f. and u.h.f. antennas, which do not usually use earth as a part of the radiating system, are usually tested in so-called free space conditions. Of course they too must be located a finite distance above earth, and much experimental work has been done to determine the effects of earth reflections on vertical patterns from such antennas. It is a well-known fact that in radar, for example, there may be "blind" zones at certain angles of elevation above the horizon, and that enemy aircraft can sneak in on a target by flying down one of these blind zones, which are due to cancellation of signal by reflections from earth or ocean surface. So, patterns determined at v.h.f. and u.h.f. are usually approximately those of free space, and the engineer must then compute or use empirical charts to determine the effect of earth reflections if he needs to know it.

To sum up,the use of models is a very effective technique and a great time-saver for the antenna designer, but it must be supplemented by practical knowledge of the performance of antennas in their true environment. Modelling is one of the engineers tools — others are actual results obtained by others in the past, texts and charts containing design information, and actual proof-of-performance measurements in the field on the full-scale "live beast"! ■

20

QUESTIONS AND ANSWERS

In this, the final chapter of the book, the author, lists and answers some of the questions he has received from readers during the series.

ONE of the things that goes with writing articles for amateur magazines is the large amount of correspondence from readers seeking advice and assistance. I have always tried to answer all letters received, albeit not as promptly as I would like to have done it, due to the pressure of other things. One thing that any author appreciates is the enclosure of a stamped, self-addressed envelope. This is only common courtesy, and it usually brings a more prompt and complete answer. The mail from my articles on vertical antennas has been considerable. In fact, it has been greater than that from anything else I have ever written, which shows that readers are interested in learning about antennas. This has been especially true of those who want to use verticals on 75 or 160 meters. The recent power increases on 160 meters and the increasing availability of s.s.b. equipment for 160 meters have led to considerable interest in antennas for that band, shown both in the correspondence and in my on-the-air contacts.

In thinking of what to say in this Chapter, I decided that it would be appropriate to give some practical information on problems which confront our readers, and what better way to do it can there be than to quote actual questions received, and the answers I have given? I am aware that this may lead to a flood of additional correspondence with questions, but I am sure that I can make some provision for handling it, and perhaps publishing the answers for the information of all. Incidentally, this has been something I have wanted to do for a long time, to get out a small book on verticals, a subject which, until now has not been covered in any depth in print.

Questions and Answers

1. Q. *I am new in amateur radio, and don't know much about antennas. What do you recommend for 80 meter c.w.?*

A. The answer to this will depend on several factors which I do not know, such as the space you have available, and what areas and distances you expect to work. If you have the space, and a pair of tall poles or other supports, you can put up a dipole cut for about 3650 kHz, feeding it in the center with a balun and coaxial line. The height above ground will determine the vertical angle of maximum radiation, in general the greater heights giving lower angles. If it is only 30 to 40 feet above ground, for example (less than ¼ wavelength), most of your radiation will be at high angles, which will give you good local area coverage within 200 or 300 miles. A vertical antenna of

¼ wave height or less, with a good ground system, will give you a lower angle of radiation, which will extend your coverage out well beyond 1000 miles reliably. The vertical will take less space than the horizontal dipole, but it does require a ground system if it is to work efficiently.

2. Q. *What would be the result of turning my three element tri-band beam on end and operating it as a rotary vertical array? Could I still mount it on my 50 foot tower and pipe mount?*
A. This would be very nice if you could mount it somehow without any steel pole, tower, or supporting structure. However, the presence of such items parallel to the elements will distort the beam pattern and one cannot predict the results. Also, there is the matter of feeding it. Unless you could devise some means of carrying the feedline out horizontally from the center of the driven element for considerable distance, you would have an unbalanced situation which would give you antenna currents on the outside of the feedline. (This would be the case if you were to bring the feedline down vertically, parallel to the elements.) It is impossible to predict what you would get from this installation without building a small model (v.h.f. or u.h.f. range) and actually measuring the patterns on a model range. I once considered using a ⅝ wave grounded base, gamma fed vertical as a driven element, mounting the rotor right in it at the proper distance above ground, mounting a crossboom above the rotor, and hanging vertical reflector and director from the crossboom. By this means one could solve the feed problem and the support problem in one stroke. However, such a thing would be a single band affair. I may yet do it as an experiment, on 10 or 15 meters, where the size would be reasonable and easy to work with.

3. Q. *Will a ground system under my horizontal beam give me improved performance? What should be its configuration?*
A. A ground system under a Yagi array is a waste of money and effort. The radiation from the array is horizontally polarized, and is mainly horizontal in direction. Therefore a ground system beneath it will not affect it. The ground (earth) which does affect it is that in the Fresnel Zone (first reflection zone) which is many yards away from your antenna. Of course this area is not usually under your control, and you are unable to do anything about its character. If you live on a body of water you will have an excellent, low loss reflecting surface in the Fresnel Zone, and you can consider yourself fortunate indeed.

4. Q. *I have a 40 foot cabin cruiser, and wish to operate mobile with my KWM-2 in it. What do you recommend for an antenna? How can I get a ground connection inasmuch as the boat has a wooden hull?*
A. I recommend you use one of the several good mobile whips now on the market, if your operation is going to be on the amateur bands only. One with traps or replaceable coils would be very good. Make sure it is one with the loading coil in the whip rather than at its base, inasmuch as this will bring your current maximum up into the antenna instead of having it appear in the base loading coil where it contributes nothing to the radiation. If you wish to operate in the ship-shore bands between 2 and 3 mHz, I would suggest one of the wound whips for this service which you will find advertised in such sources as the Radio Master, *etc.* A ground connection can be obtained by installing a bronze plate about 2 by 4 feet on the bottom of the boat, and bringing several bronze bolts through the hull (approximately caulked, of course), for connections inside the boat. The motor, fuel tank, and other associated metal parts should be connected to this ground plate. You may have to install an ignition noise suppression system also, as you would do in an automobile. By the way, don't install the plate near the bow. Put it back near the stern, so that if you go fast in your boat it will not lift out of the water and unground you!

5. Q. *I have a Hy-Tower vertical, and for a ground connection I have connected to a well casing which goes down 40 feet. The water table is at about 25 feet. Is this sufficient, or do I need some radials? If the latter, how many?*
A. You have a good d.c. and lightning protection ground, which is worthless for h.f. radio. If you will read Chapter Twelve, you will see that at h.f. the depth of penetration of the earth currents is very small, so what you have down below at 25 feet makes no difference. You need ground radials just below the surface to provide you with a low loss return path for the earth currents. Put in about 25 to 50 radials of number 16 wire, about 2 or three inches deep. This can be done with a powered lawn edger to make a slot in the sod, or by one of those half-round edger blades powered by your foot stepping on it. I did 50 radials (2000 feet of wire) in two Saturdays. When you get this ground system down and connected, you will be surprised at the increased performance of your vertical antenna.

6. Q. *I am in a house which is surrounded on three sides by other houses very close, and* *there is not much yard space. If I put a quarter wave vertical for 40 meters on top of the house, will it work? I can't put down any ground system.*

A. Yes, you can put a vertical on top of the house. For a ground plane of sorts, you can lay some radials on the roof of the house. I would lay own at least 16 of them, to assure a reasonable efficiency. They should be as long as possible. If necessary, run them over the edge of the roof and down the sides a bit. You could even connect each one to a copper ground rod driven into the earth at ground level. Make sure that you do not have any unbonded contacts to gutter pipes, *etc.,* for non-linear joints which develop at such places will give you plenty of TVI problems. Of course, you will not get the classic patterns out of this antenna, but it will work reasonably well. Feed it with coaxial line run up to its base.

7. Q. *Doesn't a vertical cause more TVI and BCI than a horizontal antenna?*

A. I don't believe it does. I have used one ever since 1959. I have had several TVI complaints which have been solved in accordance with the FCC's established program, by having the set owner get from his set manufacturer, at no charge, the necessary filter, and I have installed them as a courtesy. I found that with my Mark IV Antenna, which uses the tri-band beam on 10, 15 and 20, and the tower (ladder) as a top-loaded vertical on 40, 75 and 160, I had more complaints from the high band operation on the beam than I did with the low band operation on the vertical. This is probably due to several factors. First, at the higher frequencies, house wiring, TV antennas, speaker leads, *etc.,* are a larger fraction of a wavelength than they are at the lower frequencies, and thus they tend to pick up more h.f. energy. Also, the beam concentrates its signal, and when it was pointed at certain neighbors they really got it! On the other hand, those with outdoor antennas had lower frequency energy picked up by the long downleads. This caused a beat between the TV sound carrier and the h.f. transmitter signal, which caused a spurious signal to appear in the TV set's picture passband. The filters solved this problem also. As I recall it, when I used a simple vertical on 15 and 20 meters, before putting up the beam, the TVI problems were less severe, and I feel this difference is due to the beam's present concentration of the energy in one direction. BCI, in the little transistor sets, is impossible to eliminate, and I don't see any difference between the beam and the vertical in this respect. If there is any difference, it is probably due to frequency more than to polarization.

8. Q. *I have a crank-up tower for my beam. How can I use it as a folded unipole? Where should I tap onto it?*

A. You have a problem! You can either choose to crank it up and down, or to feed it as a folded unipole, but not both. In the first place, *you will have to bond the joints between sections of tower to assure a good connection,* first from the standpoint of a good r.f. joint, and second, from the standpoint of preventing any non-linear joints which would generate TVI. When you have done this you will not be able to crank it up and down. Also, if you do not bond it, each time you crank it up and down it is probable that the tower sections will not make connection with the same exact path length, and the tuning will change a bit. Where should you tap it? That depends on what bands you want to work. I suggest you read Chapter Eighteen on the Mark IV DX Antenna. I would go to the top of the tower with the 160 meter fold, and this could also be used for 75 meters if the tower is less than about 50 feet. If more, tap down at about the 30 foot level. For 40 meters, tap it at about 20 or 25 feet. Matching can be done with simple L networks consisting of two capacitors.

9. Q. *I tried feeding my 50 foot tower on 160 meters by tapping onto it at about 20 feet above ground with the coaxial line, but it would not match, so I gave up the idea. Now, with your articles you make me think it can be done. Where should I tap on it and how should I feed it?*

A. Well, you tried, but forgot the fact that you must have an impedance matching network in the circuit. As in the previous answer, take the fold wire all the way to the top of the tower and connect it there. Even though it is a 50 footer, the impedance looking into the fold on 160 meters will be of quite low resistance and rather high inductive reactance. This means that you have to use two capacitors in an L network to match it to RG-8/U line. Refer to Chapter Eighteen on the Mark IV for an example of this feed on 160 meters,and for typical capacitor sizes. I should mention one thing, though, and that is that your area is now allowed 1000 watts daytime on 160 meters, and the voltage ratings of the capacitors will have to be increased over those used in the Mark IV. Those had only 0.1" spacing, and when I tried 1000 watts they arced over due to the currents and voltages involved. I had a couple of vacuum variables which I put in, and now all is well. I also had to increase the contact spacing on the relay that lifted the coax braid off ground. But perhaps you will not have that problem. If you are going to run only 100 watts or so, you won't have this arcing problem with 0.1" spacing capacitors.

10. Q. *I saw sometime ago in one of the magazines an article about verticals in an array which were grounded at their bases and fed at their tops. Why have you not covered this in your series? Where can I get information on this feed method?*

A. You have a good memory. That item appeared in QST for April 1959 in an article by R. W. Johnson, W6MUR, entitled, "The Groundpole Antenna." I did not cover it in my series because the configuration is not purely vertical, but also has horizontal elements which radiate. Basically, the idea is as follows. Take two ¼ wave verticals, for example, separate them ½ wavelength, ground their bases and feed them by means of a center-fed half wave dipole whose ends are connected to the tops of the vertical elements. The radiation pattern from this combination will be complex, and would require computer analysis, or actual scale modelling for its determination. There are other configurations which may be used also, involving more than two elements. I should emphasize even more than W6MUR did that a good ground system is necessary for this type to work efficiently. In fact, inasmuch as maximum current occurs at the grounded ends of the vertical elements, ground currents will be quite high, because the whole thing can be considered as a closed loop with earth forming one part of the loop. It follows that this return path should be as low loss as possible, and even if many radials cannot be put down, it is essential that there be a low resistance ground bus between the bottom ends of all vertical elements. This can be made of copper strap or several number 16 or number 12 wires run together in parallel. You might consider this type of antenna this way: normally, vertical arrays are fed from the base of the elements by a non-radiating system of transmission lines. In this case, the vertical elements are fed from the top with single wire transmission lines that are allowed to radiate and contribute to the pattern, which then becomes complex and requires considerable mathematical analysis for its determination. My congratulations to W6MUR for his fine article.

11. Q. *Will my Quad work as top loading on my tower for 75 and 160 meters, as your triband beam does on your Mark IV DX antenna?*

A. I see no reason why not if it has a metal boom and supporting arms. These plus the Quad element wires would constitute a considerable added top capacity on the tower, and it should work fine. I don't have a Quad but common sense tells me that it will work.

12. Q. *In your Mark IV DX antenna I notice that you have your CDR rotor motor physically and electrically in series with the vertical element of the antenna, with no bonding across it. Doesn't the motor winding burn up? Are you sure of the r.f. path through the motor housing? How do you protect it?*

A. Before making this installation I asked several CDR representatives whom I met at a local electronics show whether they had any reservations about the idea. They felt that the large number of ball bearings in the raceway would provide a good multiple contact between upper and lower portions of the housing. They did recommend that the rotor control cable conductors be bypassed to ground at the base of the antenna, inasmuch as the control cable is in effect in parallel with the tower itself r.f.-wise and will carry some r.f. current. I did this. Each of the seven conductors is bypassed to ground at the base of the tower with a 0.01 mf disc ceramic capacitor. These capacitors are housed in an aluminum beer can (empty), entrance to which is through the bottom end to keep water out. The can is mounted on one of the tower legs. I do notice a very slight upward swing of the rotor indicator needle on 75 meters when I talk, but it is less than ¼", and I attribute it to r.f. getting into the cable somewhere in the shack itself. Incidentally, there *is* an r.f. path *across* the rotor motor housing, and this is through the outer conductor of the coaxial line and through the balun coil. Obviously this is also a d.c. path for static discharges.

13. Q. *In finding the gain of a directional array, shouldn't you take the r.m.s. of the whole pattern through the vertical angles, instead of just the r.m.s. of the horizontal pattern?*

A. As I have said before, *gain is meaningful only when referred to a specific direction (horizontal and vertical angle).* It is standard practice among consulting engineers to use the horizontal pattern (vertical angle zero degrees) because that is the easiest one to measure by walking or driving to measuring points. Also, consultants are interested in ground wave coverage, usually. In some cases it is necessary to know the skywave pattern at a particular horizontal azimuth angle. The FCC insists on knowing the predicted vertical patterns, taken every 10° vertically, for all horizontal azimuth angles. If one wishes to know the gain of an array at a particular vertical angle, he can compute it by putting the vertical angle into the pattern equation, computing the pattern at that vertical angle, finding its r.m.s., and computing the gain, as I did for the horizontal plots in Chapters Nine and Ten. I did not do it for the vertical angles because I felt it would add too much complication to the articles for the average reader to follow. However, it can be done by anyone so deeply interested. This item came from a college professor who had worked out a computer program for deriving the whole three-dimensional pattern of an array, and he felt that to know the gain of the array one should base it on the r.m.s. of the three-dimensional pattern. It is very commendable that he has worked this out. Consulting engineers now use computers exclusively for array computations. However, I reiterate that gain is meaningful to the communication system user only when referred to his specific direction of interest. The gain at other directions is nice to know because he can then tell whether his antenna is wasting power in spurious lobes, but it is not essential to the basic question of gain in the desired direction. And in my case I chose the horizontal plane for ease in measurement.

14. Q. *In your directional computations you show only the horizontal patterns. What about the vertical angles? These antennas can't be as perfect as you have shown them. Isn't there power lost in high angle skywave lobes?*
A. Yes, there is power in skywave lobes, but in the case of the simple two and three tower patterns shown, the skywave lobes are small, relatively. In the case of in-line arrays, one can generalize and say that in each null of the (horizontal groundwave) pattern there will be a small skywave lobe. Its angle above the horizontal will depend on several parameters of the array, the most important of which is the factor f (θ) for the tower heights involved. In the case of the arrays of other configurations and more elements, such as parallelograms or "dog-legs," the skywave lobes can be found by computer. I know of one consultant who designed a multi-element array of irregular configuration for a client some years ago, who did not compute the whole pattern at 10° vertical angle increments (before it was required). When the array was built, he could not make his horizontal r.m.s. come up to the value required by the FCC. In other words, he was losing too much power in skywave lobes and in circulating power within the array. It was a "losser." The array had to be torn down and redesigned. Naturally the client became quite irate,instituted a law suit, *etc.* This was quite embarrassing as well as costly to all concerned. So, the answer is, "Yes, there are skywave lobes, but in the simple patterns they are minimal."

15. Q. *My radio room is on the third floor of our house. I would like to use a vertical on the roof, but how do I make a ground connection from this height?*
A. Obviously you cannot make an r.f. ground connection to earth from this height, because your ground lead will be a part of the radiating system and will place your entire installation at some r.f. potential above ground, besides introducing possible losses from coupling to house wiring, plumbing, *etc.* I suggest you install a *ground plane* antenna on your roof, feeding it with coaxial line, in the same manner as I suggested to the writer of question 6.

16. Q. *There is a power line which runs along the rear property line of my yard, and these poles carry not only power lines but also the telephone cable and a CATV cable. Drops from these services come into my house at both sides of my yard, and thus my backyard is surrounded by unwanted conductors. I do not want to erect an antenna in front of the house, but what can I do in the rear? There is not room to put up a horizontal wire without bending it several ways. The yard itself is fairly open and clear.*
A. This was my problem when I moved here to Thousand Oaks. In Maryland we had a large yard, larger than most, and the rear portion was completely clear of all obstructions. The power lines, etc. were on the street, in front of the house. Thus the Mark IV had a very clear field in which to perform. In house hunting here I had to stay away from new development areas where all utilities are underground and not even a TV antenna is permitted to be erected outside because it would "spoil the environment"! I had to settle for a piece of property which is fairly clear in itself, but is surrounded as mentioned in the question. (I am not alone in this, as most amateurs here have the same situation.) The Mark IV was the answer to the problem, in that it could be erected in the center of the backyard, with clearance all around it. The power line is about 40 feet away, to the rear. The drops come along the sides of the lot and enter the eastern end of the house. On our left our neighbors' drops do the same thing. With the Mark IV I note no degradation of performance nor any r.f. getting into the lines. I would like to get the Mark IV's beam up higher to get further above

the power line as several of my amateur neighbors have done. I also put up a horizontal wire, hung from the top of the Mark IV tower from a 20 foot dacron rope, with its other end on a short stub mast on the front of the house. Its total length including downlead is about 90 feet. I use it against ground, with a base tuning unit which is in the radio room, mounted near the entrance of the ground cable from the Mark IV's ground system and connected to it. This long wire is used for frequencies outside the amateur bands, and also for v.l.f., l.f., m.f., and h.f. receiving with my R-389 and R-390 receivers. When I use it for transmitting on 3.9 or 7 mHz, the power line drop into the house becomes excited by r.f. and is "hot". So is the telephone line! This is due to proximity to the long wire and the horizontal polarization of the long wire. There is no such pickup from the Mark IV vertical which I can notice. I can also tune and use the wire for 160 meters, but I do not note any pickup from it to power and telephone lines on that band. The line length seems to resonate on 3.9 and 7 mHz, and energy from the horizontal wire couples in readily. Another important factor is that the r.f. feed to the Mark IV is by means of the buried coaxial line, whereas the feed to the wire antenna passes thru the wall of the radio room via a large lead-in bowl insulator, and thus the hot end of the antenna is at the house. This can't be avoided in this case because I needed to be able to tune the wire antenna for several different frequencies not related and could not use a dipole fed with transmission line. End feed was a necessity. Thus, for the situation you describe, the Mark IV is a good answer, where you do not have the space for a dipole and are surrounded by unwanted wires, cables, etc. The vertical is the only thing which will fit.

17. Q. *Can I mount a v.h.f. array on top of the Mark IV? If so, how far up should it be above the Mark IV's beam, and what do I do with its coaxial feed line? Will it adversely affect the Mark IV, or vice versa?*
A. There are several such installations here in my area, due to the popularity of 2 meter f.m. The v.h.f. array is usually mounted above the h.f. beam at a distance of six to ten feet. The feed line may be brought down the pipe support and on down the tower with sufficient loop across the rotator to allow rotation. I suggest that a junction made of a PL-275 adapter and two PL-259 plugs be used up at the top of the tower, so that the v.h.f. beam may be removed from the tower if desired, without removing all the coaxial line all the way down to accomplish it. This junction should be waterproofed with vinyl tape in the same manner as I did the one in the h.f. coaxial line of the Mark IV. If I were going to do this, in order to lighten the bending moment on the CDR rotator, I would mount the h.f. beam right on top of the rotator, and extend its pipe support on up for the six to ten feet desired for the v.h.f. array. Or a lighter pipe support can be used for the v.h.f. array with a suitable coupling to connect it to the pipe which mounts the h.f. beam to the rotator. This would lower the h.f. beam by about four feet on the present Mark IV, but many of you will be doing this with higher towers and that four feet will not matter too much in h.f. performance. There is no need to wind up the v.h.f. coaxial line in the form of a choke. That was done for the h.f. line of the Mark IV to eliminate the necessity of using a balun, not for reasons of isolating the beam from the tower. With the v.h.f. array mounted on top, the total height of the Mark IV used as a vertical radiator on 7, 3.9 and 2 mHz will be increased by the height of the pipe support for the v.h.f. array. This will be a fortuitous gain in performance of the vertical radiator, and this added height should be figured into any computations of vertical radiator height vs pattern performance.

18. Q. *What if I want to use the guys of the Mark IV tower as top-loading for the tower used as a vertical radiator? Will this increase the effective height and enable me to get a better impedance match to the "fold" wire on 160 meters?*
A. Yes, it will, but how much it will improve performance cannot be calculated exactly, inasmuch as there is already considerable top-loading in the form of the h.f. beam on top of the tower. If you will refer to Chapter Four of this book you will see the results of studies made by several engineers on scale models and actual full size v.l.f. and l.f. towers, and you will find design data for the top-loading by means of using portion of the guys. I suggest you follow this information if at all possible.

19. Q. *In a past article you gave information on the NORD antenna being used for l.f. by the Navy. Can you give me any more information on this? Is it still in use?*
A. If you will refer to Chapter Four of this book you will see the NORD antenna with its design data. The NORDS are still in use and are carrying the multichannel transmissions of the Pacific Fleet Broadcast. Basically the NORD consists of three closely coupled tuned loops using ground return to complete the

circuits of each. They are fed in common by feeding a vertical radiator as a folded unipole. The advantage of the NORD is that it may be used on a tower which is electrically quite short at the frequency involved, with a reasonable efficiency. A series fed tower of the same height must be fed through a large loading coil or helix, inasmuch as its base impedance has high capacitive reactance. The input impedance of the NORD's fold is inductive, and it can be matched to coaxial line by means of a low-loss network of vacuum capacitors. The helix of the series fed tower, on the other hand, has considerable ohmic loss, which at l.f. or v.l.f. can cause lower overall efficiency. Note the plot of the various components of total antenna system resistance of a v.l.f. antenna shown in Fig. 24 of the Chapter Four. We certainly don't want any more ohmic loss in the system than we can help having.

Another factor in favor of the NORD vs a series fed tower which is top-loaded by its guys at v.l.f. or l.f. is that with high power there are corona problems at the ends of the top-loading guys. For powers of 50 and 100 kw which are not uncommon at l.f., the guy insulators which isolate the top-loading from the rest of the guys go into corona and break down. This has been a serious problem. With the NORD no such problem exists, because there are no high potential points like that. Instead, the NORD must be designed to carry considerable current in the guys and their tuning capacitors. These capacitors must be rated for the currents involved.

Incidentally, I was consultant on the early experiments with the NORD. We obtained special FCC authority to test using a 35 foot vertical radiator on an unused frequency in the standard broadcast band in the Maryland area, and made actual field strength measurements of various configurations of NORDS, and learned much about its ability to be tuned with various bandwidths *vs* efficiency, and tower heights. A ground system of a large number of radials was actually laid down in a pasture in the suburbs. It was a very informative study.

20. Q. *For a vertical antenna, do I need to be at a high location for good performance on 2, 3.9 or 7 mHz? I note that many standard broadcast stations are located in swampy, ground. Is this required at h.f. as well as at m.f.?*
A. Yes, this type of location is certainly a big help at the frequencies you mentioned. Of course your XYL may not want to live in a swamp! But if you will refer to Chapter Two and look at Figures 7, 8 and 9, you will see the results of locating in an area of poor ground conductivity *vs* an area of sea water. The latter contributes much to efficiency, especially at h.f. Height of terrain in itself is good, but you don't often find *high,* wet, swampy ground! In the early days of broadcasting it was thought that m.f. stations should be on hills or mountains — I recall several, such as the old WHAM, Rochester site at Victor, N.Y., and WTIC's site on Avon Mountain near Hartford. But as more studies were made of propagation it was found that the groundwave coverage which was desired was better from the low, wet sites than from high hills. In the case of h.f., the wet site with good ground conductivity will improve the low angle radiation by reducing earth losses. Now if you are considering only propagation on the higher bands such as 14, 21 and 28 mHz for which I used a Yagi beam on the Mark IV, then a good height, in the clear, is much to be desired. A mountain top is a beautiful site for a Yagi beam or a Quad! They do not use ground systems nor the resulting low angle propagation. But even with them, a site on water is very desirable, for the reason that the first reflection zone of the horizontally polarized radiation from such an antenna will fall out on the water, and there will not be as much loss as there is when the reflection zone falls on earth. A ground system under a horizontal antenna is useless because it cannot be extensive enough to reach out to this reflection zone.

21. Q. *I understand that f.m. broadcast stations use vertical antennas now. Is this correct, and if so, why is this?*
A. You are only partially correct. F.m. broadcasters are now permitted by the FCC to use *both* horizontal and vertical polarization in their transmitting antennas. *They cannot use exclusively vertical* polarization, and *they may use only horizontal* if they wish. But most of them are using both, and the reason for it is one of propagation and polarization. More and more automotible radios are now equipped to receive f.m. as well as a.m., and many people enjoy f.m. stereo reception in their cars. For the reason that many auto radio antennas are short vertical whips, and for the reason that in moving about on the streets and highways on encounters zones where polarizaton of an f.m. signal has been shifted about by reflections, it is advantageous to transmit both polarizations and thus give the automobile user the best possible reception of f.m. It is a form of polarization diversity.

22. Q. *I have been told that when one is using a vertical antenna for transmitting, the signal can be received well only on another vertical antenna. Why is this?*
A. Your informant forgot to mention the frequency band being used for this. I assume he meant the h.f. bands, over long distances. His statement is false when applied to the h.f. bands. This is because in h.f. long distance transmission where the signal has undergone an ionospheric reflection, the wave is no longer vertically polarized, but is polarized in a random and varying manner due to the ionosphere. The polarization of the transmitting antenna makes not one bit of difference as far as received polarization is concerned. In fact, we can take advantage of this shifting polarization to improve h.f. reception by use of *polarization diversity.* I do it here in receiving h.f. broadcast such as BBC, Radio Nederland, and others, by using a diversity connection of my R-390 and R-390A receivers, with the horizontal wire on one of them and the vertical (Mark IV) on the other. It is very seldom that both polarizations fade down at the same time. With a suitable combiner which ties the receiver a.v.c. busses together, the set with the stronger signal will bias down the one with the weaker signal, and a fairly steady, noise-free output results. It is much better than one alone. Now, to return to the question — for v.h.f. or u.h.f., where we are using *line-of-sight* propagation, the transmitted polarization requires a receiving antenna of like polarization, *provided the transmission path is stable and free from sources of reflections* which vary or shift the polarization. See the answer to the previous question for an application of this to f.m. reception at v.h.f. If the path is varying, or reflections are changing the polarization of the received signal, we can use polarization diversity at v.h.f. or u.h.f. by using two receiving antennas of differing polarization. (In the f.m. broadcast case the diversity is put into the transmitting antenna instead.)

23. Q. *My horizontal dipole is quite frequency sensitive. I am center-feeding it, and the v.s.w.r. at the ends of the band rises to high values whereas it is well-matched at the middle of the band. What can I do? Will a matching unit at the transmitter help to lower the v.s.w.r. of this antenna?*
A. No, it will not. That is a popular misconception. It will lower the v.s.w.r. of the load which the line presents to the transmitter, but the same high v.s.w.r. will be present in the line itself and in the antenna which is the load on the far end of that line. *The matching unit will serve only to make the transmitter final stage happier.* The only way you can improve the antenna's v.s.w.r. is to broadband it. I assume you are using a single horizontal wire. The wire is of quite small diameter when compared with its length. Thus it is a very high-Q device. Refer to Chapter Three for a discussion of antenna bandwidth and K_a These principles apply to a horizontal dipole as well as to a vertical monopole. A wire antenna may be broadbanded by making its diameter greater in relation to its length. A "fat" antenna made of a cage of wires (like some of the old-time l.f. or m.f. antennas) will have a much better bandwidth than a slim single wire. Simply apply the information shown in the references by Schelkuoff[6] and Laport[7] to the broadbanding problem. Two fans of wires may also be used in lieu of a cage dipole. The fans are fed at their apexes. Various broadband types are discussed in Chapters Seven and Eight. The same principles will apply to dipoles, as to the verticals.

24. Q. I have noted that some standard broadcast directional arrays employ towers of differing heights. Why is this? I would think they would have to be all of the same height.
A. The elements of a vertical array which are base tuned can be of differing heights. When this is noted, you will often find that one of the towers (the tall one) is ⅝ (0.625) wavelength tall. Reference to Fig. 6 in Chapter Two will show the reason for this — that height gives the greatest groundwave field. For reasons of economy, where possible the array is designed to use a shorter tower or towers to put the nulls into the directional pattern by the proper spacing and phasing of them with reference to the main tower. Also, perhaps the taller tower is desired to support an f.m. or TV broadcast antenna for which height is desired.

25. Q *When a series fed vertical radiator is used, how can one bring tower lighting circuits, v.h.f. antenna feed, etc. across the base insulator? Isn't it shorted out by such an arrangement?*
A. It would be if suitable precautions were not taken. In the case of broadcast towers, large chokes are used in the tower lighting circuit to carry the a.c. power but to prevent shorting the base insulator. There is another type of device called an Austin Transformer which is also used, but this is quite expensive. It is a large air-core transformer which isolates the a.c. mains on the ground from the a.c. line on the tower. With regard to v.h.f. coaxial lines, there is one way, and that is to coil up small size coaxial line in form of a large choke coil on a suitable insulating form or spreaders, and

when this is done it will pass the v.h.f. but not short out the tower base insulator at l.f., m.f. or h.f. In the case of a single frequency vertical such as a broadcast tower, the v.h.f. coaxial line of larger sizes may be isolated from the tower for a distance of ¼ wavelength at the broadcast frequency before it is grounded and comes into the building, thereby forming a quarter wave stub at the broadcast frequency. The open end of the quarter wave stub is across the base insulator, and at the broadcast frequency it acts like it isn't there. This is rather expensive if much additional heavy copper coaxial line is required. A better and cheaper way of doing it, which is foolproof, is to convert the vertical series fed radiator to a grounded base folded unipole. Then all these added circuits, such as tower lighting, v.h.f., f.m. or TV lines, two-way radio lines, CATV lines, etc. can be brought down the tower and buried underground because there is no base insulator to worry about. At the main transmitting frequency the only thing hot with r.f. is the end of the fold (feed wire), and that is connected only to the matching unit which can be a pair of capacitors. Refer to Chapter Eighteen on my Mark IV. The coaxial line for the beam, the rotator control cable, and anything else I might ever want to hang on that tower, are simply brought down the tower to ground, for there is no base insulator to worry about. This is what we in consulting do when an a.m. station wishes to add f.m. The a.m. tower is converted to a folded unipole, retuned and used as before, with the f.m. antenna mounted on top of it, and its feed line coming down the tower and into the building, usually underground. ■

INDEX

E

F

G

Q

R

S

T

V

ADDENDUM

Antenna Design For 160 Meters

SINCE the original publication of this book, there have been many advances in methods of design of antennas. Computers have come into use, and now most directional antenna designs like those in Chapters 5 and 9 are done by computer. Vertical radiation patterns of single vertical radiators and arrays are plotted by computer. Much time consuming and tiresome calculation is thereby eliminated. The computer can be given a set of parameters which must be met, and allowed to search for the solution or solutions. There have been several articles in the magazines on the use of computers in antenna design.

Also, our Federal Communications Commission has been gradually opening up the 160 meter band for increased amateur use, as Loran A is being phased out in various parts of the country. Interest in that band by amateurs is increasing. Many have asked me on the air and by mail, "What is a good antenna for 160 meters?" Attempts to refer them to various designs in this book have been unproductive, as the book has been out of print for a few years. With that fact in mind, the publisher and I have combined efforts to not only reprint the basic text with minor corrections, but also to provide this Addendum with information on suitable antennas for 160 meters. Much of what is said here will refer to the various designs shown in the book, in detail, with discussion of their advantages and disadvantages.

The band 1.8 to 2 MHz is a good one for short distance communications at most times during the year, and at certain times it is useful for fairly long distances. When located in California in the early '70's, I did contact a friend in New York State on voice on that band, and also several in Ohio and other midwestern states. I was using the Mark IV antenna described in Chapter 18 of this book. Now here in Florida on the "Space Coast", I have not put up the full height of the Mark IV due to height restrictions, but have limited myself to 30 feet, using an aluminum ladder as before. Use of 1.8 - 2 MHz has been with a long wire, but I know that a vertical would produce better results. When time permits I plan to do some experimenting with various configurations of short vertical antennas using the 30 foot tower (ladder with beam on it).

In this Addendum I want to summarize design information for 1.8 - 2 MHz antennas, for those who are interested. First, let me make some basic comments.

Use of a baseloading coil, or even a tuning unit by itself, is a poor method of achieving an impedance match to a short antenna, for it puts the current loop in the coil or tuning unit, instead of up in the antenna itself where it can produce radiation. Top-loading, or top-loading over a loading coil, is the best method of design for a short vertical radiator, producing the greatest efficiency.

This is because the input resistance is increased by such action, and also the input reactance is favorably reduced, making a match easier. The use of the folded unipole feed method is even better as it enables us to use a capacitive matching network, thus eliminating coil loss. The input of a folded unipole is always inductive in nature.

Also, a very important thing to remember is that a good ground system is a necessity when operating on 1.8 -2 MHz. This is especially true in soil like that which we have here, which is very sandy and of poor conductivity. Unless the water table is very close to the surface, consider it to be poor soil! One broadcast station here in Cocoa has its tower on a pier in the Indian River, with the radials buried in the mud of the river bottom, thus providing an excellent salt water ground system. Chapter 2 of this book covers such factors as numbers of radials and length of radials vs antenna efficiency. Figures 14 and 15 show these factors very well, and should be used as guides in designing a ground system. If you can't lay out 1/4 wavelength radials on your property, run them to the limits of the property, and add more of them close in.

When I added ground radials to increase the number from 16 to 50 in Maryland, the improvement was very noticeable with my Mark IV antenna, described in this book, Chapter 18.

By ground system I do not mean connection to a buried water pipe, or to a well casing. I mean *radials* which extend out from the base of the antenna for at least 1/4 wavelength if possible, within the property limits. If that length is not achievable, then do the best you can. If the radials are short on one side of the plot, then add wires to help compensate for that. Radials should be buried about 6 to 8 inches deep. Putting them deeper will increase earth loss.

There is also another type of ground system which should be mentioned here, and that is the counterpoise, or above-ground ground plane. This is covered in Ed Laport's very fine antenna text.[37] Incidentally, I had the privilege of working with Ed many years ago at RCA International Division. He was truly one of the experts in his field. This type of ground plane is also discussed in two excellent magazine articles by Doty.[66, 67] Unfortunately, the author does not tell under what organization or direction the tests described were made. They are simply called "the Fletcher tests". Such information would have lent prestige and additional credence to his excellent reports.

The usual counterpoise system consists of radials, hopefully at least 1/4 wave long, supported above ground by poles. They act as a ground screen, collecting the currents which would normally go into the earth to buried radials, thereby lowering earth losses in the immediate vicinity of the tower base. Tests indicate that the radials of the counterpoise are capacitively coupled to the earth beneath them. Doty went even further, and tested a system with insulated radials laid directly on earth. By his tests he showed that such a ground plane can increase the efficiency of a vertical antenna system over that which would be expected with a system of buried radials. This is to be expected since earth around the antenna, which is lossy, is not in the picture. It is for this reason that at VLF and LF such systems have been used[37] to reduce the losses in the earth around the tower, which at very high powers can be considerable. These articles are recommended reading.

There is one thing which cannot be overlooked, however, and that is that earth currents are introduced in the earth for a distance of several wavelengths from the tower, and provision should be made to convey them back to the system. Each radial of the counterpoise or ground plane screen should be terminated in a copper ground rod, driven to a depth of at least six feet.

There is one serious problem with this type of ground plane, and that is that it requires a backyard or clear space of considerable dimensions for use on 160 meters, something which most amateurs do not have, and which most XYLs would not permit to be installed. Burying the radials gets them out of the way, and makes the system more "palatable" to the XYL.

Doty makes the statement that such a system, with a 30 foot vertical radiator and a 100 X 100 foot ground screen, would be a very efficient antenna for 160 meters.[66] Theoretically this is true, if the antenna can be made to take power. An electrically short antenna has an input impedance of very low resistance and a very high capacitive reactance which must be cancelled out by a helix or loading coil at the base, or by some other method of feed than series feed. He does not say how he would propose to feed such a short antenna. This leads us into the next portion of this Appendix which deals with electrically short antennas and methods of feeding them efficiently.

As most readers will not be able to erect a 1/4 wave vertical for 1.8 - 2 MHz, electrically short antennas must be used, and suitable means must be used to make them as efficient as possible. There are several types covered in this book, and most of them involve the use of top-loading a vertical radiator to increase its efficiency by increasing its radiation resistance and reducing its losses.

The first of these which I want to mention is the top-loaded folded unipole which is covered in fig. 25 of Chapter 4. Type C, with 6 top-loading guys, would be of reasonable dimensions for amateur use. Use of a grounded base tower and a fold feed will eliminate the base loading coil which is necessary for series feed, and enable the system to be matched to 50 ohm line by low loss capacitors, thereby improving efficiency.

Also in Chapter 4 is the NORD, which is a variation of the top-loaded design. This is a bit more time-consuming to adjust but it has worked well in Naval LF service. Its adjustment can provide a tradeoff of bandwidth vs. efficiency, by

---

[66]Doty, Archibald C. Jr. *et al*, "Efficient Ground Systems for Vertical Antennas", *QST*, February 1983, p.20-25.

[67]Doty, Arch, K8CFU, "Improving Vertical Antenna Efficiency", *CQ*, April 1984, p.24-31.

experiment. See fig. 29 and the text adjacent to it on page 33 for dimensions of a NORD for 2 - 4 MHz. It is of reasonable size for amateur use.

Let's move on to Chapter 7, wherein the broadband discage is described in detail. Almost every large ship in the Navy now has a discage on board for multiple transmitter use. USS *New Jersey* has one on here bow, and *Iowa, Missouri* and *Wisconsin* will each have one when recommissioned. This is the AS-2802/SRC shipboard antenna[68]. It is a very sturdy and efficient antenna. Some shore transmitting stations also have them, and at shore stations they can be made larger for lower frequency use than on shipboard. The dimensions of a discage for 2 MHz are shown in fig. 65. For a low limit of 1.8 MHz, dimensions should be increased by 10%. This is an antenna which does not take much lateral space - only vertical height. It can be fed by a 50 ohm line, and a v.h.f. or u.h.f. antenna can be top-mounted on it as described in the text, just as we do it in the Navy.

Now let's look at the Mark IV. Several people have written articles about this type in recent magazines copying its design, but this is the *original* from which their designs were taken. Chapter 18 covers the Mark IV in detail. I brought the feedline of the beam down the side of the tower (ladder) on standoff insulators, to provide a folded unipole type feed. The 3 element HyGain TH4Mk2 beam provided top-loading. The beam top-loaded the tower so that the system was a bit over 1/8 wave high (electrically) at 2 MHz, about 5/8 wave at 7 MHz, and 1/4 wave at 3.350 MHz. This is not bad at all for a 40 foot vertical radiator!

Efficiency was quite good at 1.8 - 2 MHz. This is due to the fact that I used 50 copper wire radials for a ground system, and took great care to make everything as low loss as possible. Radials were soldered to a 2 inch wide copper ground strap which connected to the tuning unit box.

This antenna is of such configuration that it can be used by almost any amateur who has a tower and a beam for the higher frequency bands. It is easy to build and easy to tune. It will provide capability of operation with one structure on the 160, 75, 40, 20, 15 and 10 meter bands, taking up no more backyard space than a tower and beam. I used this antenna for about 15 years in both Maryland and California with excellent results. The 160 meter enthusiast could do well to copy it in detail. But remember, an adequate ground system is an absolute necessity! You won't have good results, especially on 160 meters, without it. Be generous with the copper radials. Make them as numerous and as long as possible. Also, the ground bus between the tower and tuning unit should be heavy and low loss. This is mentioned in Chapter 3, on page 27.

In that connection, and before closing this Addendum, let me suggest that you read and study Chapter 12, on the effects of *real earth* on antenna system efficiency and radiation. Theoretical patterns from a computer, assuming perfect earth or a copper sheet for ground plane, are fine for laboratory experiments, but in real life, lossy earth has a great effect on vertical antenna performance. The return currents from any antenna which is close to earth must have a low loss return path to the antenna. Earth resistance, or its inverse, conductivity, have their greatest effect in the area close to the base of the antenna. This is the reason for using a buried ground system of radial wires, to provide as low a loss in the return path as possible. One author of sorts wrote many pages to me claiming that one does not need a ground plane around a 1/2 wave vertical radiator, for the base current is theoretically zero. He said the ZL's could hear him fine without it! Great! But his words do not repeal the laws of physics. The electric field from any vertical radiator induces currents in the earth, and a return path must be provided for them, for maximum efficiency. In the case of the 1/2 wave vertical, the return path should be through a tuned tank circuit between the base of the antenna and the ground system. This tank circuit can also be used for coupling into the antenna from the feed line. Figures 104, 106, 107, and 108 show the current distributions and earth lossed for both 1/4 wave and 1/2 wave antennas over earth. The references by Dr. G.H. Brown[5,52,53,] of RCA Laboratories, one of the pioneers in this business, are especially good in this regard. To summarize, for the average amateur, the poorer the earth, the more necessary it is to have an adequate ground system. *Copper* wire is necessary. Aluminum will corrode and turn to white powder in a short time in areas like this where the soil is a bit salty! My first attempt in Maryland resulted in that, and I was not near the sea at all! That white powder is not a good conductor at all!

In general, regarding this book, it has sufficient references in it to be of value to readers for a long time. I believe in references, for they lead the reader to other sources of valuable information. That is their purpose.

---

[68]Law, Preston E., *Shipboard Antennas*, Artech House, Inc., p.38, 133-141.